Workbook Solutions 6A & B

DIMENSIONS MATH 6

Singapore Math Inc.

Published by Singapore Math Inc.

19535 SW 129th Avenue
Tualatin, OR 97062
www.singaporemath.com

Dimensions Math® Workbook Solutions 6A & B
ISBN 978-1-947226-57-9

First published 2019
Reprinted 2020, 2021

Copyright © 2017 by Singapore Math Inc.

All rights reserved. This book or any portion thereof may not be reproduced or used in any manner whatsoever without the express written permission of the publisher.

Printed in China

Contents

Chapter 1
Whole Numbers

1.1 Order of Operations
    A. Expressions and Equations 1
    B. Exponents 1
    C. Order of Operations without Parentheses 2
    D. Order of Operations with Parentheses 3
1.2 Factors and Multiples
    A. Factors 4
    B. Multiples 4
1.3 Multiplication
    A. Properties of Multiplication 5
    B. Mental Multiplication 5
1.4 Division
    A. Properties of Division 7
    B. Mental Division 7

Chapter 2
Fractions

2.1 Multiplication of Fractions
    A. Multiplication of a Proper Fraction by a Whole Number 9
    B. Multiplication of a Proper Fraction by a Fraction 11
    C. Multiplication of an Improper Fraction or a Mixed Number by a Whole Number 12
    D. Multiplication of an Improper Fraction or a Mixed Number by a Fraction 13
2.2 Division of Fractions
    A. Division of a Whole Number by a Fraction 14
    B. Division of a Fraction by a Whole Number 16
    C. Division of a Fraction by a Fraction 17

Chapter 3
Decimals

3.1 Addition and Subtraction of Decimals
    A. Structure of Decimals 21
    B. Adding and Subtracting Decimals 21

3.2 Multiplication of Decimals
    A. Decimal Number System 22
    B. Multiplier and Product 23
    C. Multiplying Decimals 23

3.3 Division of Decimals
    A. Divisor and Quotient 24
    B. Dividing Decimals 24

3.4 Metric Measurements and Decimals 25

Chapter 4
Negative Numbers

4.1 Positive and Negative Numbers 26

4.2 Comparing Positive and Negative Numbers
    A. The Number Line 27
    B. Absolute Value 27

Chapter 5
Ratios

5.1 Ratios and Equivalent Ratios
    A. Finding Ratio 29
    B. Equivalent Ratios 30

5.2 Ratios and Fractions 31

Chapter 6
Rate

6.1 Average and Rate 33
6.2 Unit Rate 34
6.3 Speed 36

Chapter 7
Percent

7.1 Meaning of Percent 37
7.2 Percentage of a Quantity 38

Chapter 8
Algebraic Expressions

8.1 Writing and Evaluating Algebraic Expressions
    A. Use of Letters 41
    B. Evaluating Algebraic Expressions 41
    C. Word Problems 43
8.2 Simplifying Algebraic Expressions 44

Chapter 9
Equations and Inequalities

9.1 Equations
    A. Algebraic Equations 46
    B. Balancing Equations 47
9.2 Inequalities
    A. Algebraic Inequalities 54
    B. Graphing Inequalities Using a Number Line 54

Chapter 10
Coordinates and Graphs

10.1 The Coordinate Plane 56
10.2 Distance between Coordinate Pairs 57
10.3 Changes in Quantities
    A. Independent and Dependent Variables 60
    B. Representing Relationship between Variables 60
    C. Observing Relations between Variables with Graphs ... 61

Chapter 11
Area of Plane Figures

11.1 Area of Rectangles and Parallelograms ... 64
11.2 Area of Triangles
    A. Finding Area of a Triangle ... 65
    B. Areas Involving Parallelograms and Triangles ... 66
11.3 Area of Trapezoids ... 67

Chapter 12
Volume and Surface Area of Solids

12.1 Volume of Rectangular Prisms
    A. Cubes and Cuboids ... 69
    B. Volume of Liquids ... 70
12.2 Surface Area of Prisms
    A. Surface Area of Rectangular Prisms ... 72
    B. Surface Area of Triangular Prisms ... 74
    C. Metric Conversions and Volume ... 75

Chapter 13
Displaying and Comparing Data

13.1 Statistical Variability
    AB. Statistical Questions & Measures of Center ... 77
13.2 Displaying Numerical Data
    A. Dot Plots ... 79
    B. Histograms ... 80
13.3 Measures of Variability and Box Plots
    A. Range ... 82
    B. Mean Absolute Deviation ... 82
    C. Interquartile Range ... 84
    D. Box Plot ... 86

6A Solutions

Chapter 1: Whole Numbers
Order of Operations

1.1A Expressions and Equations

Basics

1. (a) expression (b) equation
   (c) expression (d) expression
   (e) equation

2. (a) sum (b) difference
   (c) product (d) quotient

Practice

3. (a) 3 (b) 2
   (c) 1

4. (a) $3 \times 12 - 5 = 31$
   (b) $8 \times 1 + 2 = 10$
   (c) Answers may vary.
   Example answers:
   $8 \times 2 + 1 \times 3 = 19$
   $2 \times 8 + 1 + 1 + 1 = 19$

5. Answers will vary.
   Example: $24 \times 3 - 5$
   Mr. Ivanov bought 3 crates of oranges. In each crate there were 24 oranges. He threw away 5 oranges because they were rotten. How many oranges did Mr. Ivanov have left?
   Answer: He had 67 left.

Challenge

6. (a) Answers will vary.
   Example: Ella received 15 boxes of books. There were a dozen books in each box. She put the books equally on 3 empty shelves. Later that day, she added 4 more books to the top shelf. How many books were on the top shelf?
   Answer: 64 books were on the top shelf.
   (b) Answers will vary.
   Example: There were 15 sheets of stickers. Each sheet had 12 stickers. These stickers were shared equally among 3 boys and 6 girls. How many stickers did each child get?
   Answer: Each child got 20 stickers.

1.1B Exponents

Basics

7. Base, exponent

8. 4
   2, 2, 2
   4, 2, 2, 2, 2
   2, 5, 2, 2, 2, 2, 2, 32

9. 1, 1, 1, 1, 1
   10, 10, 10, 1,000
   100, 100, 10,000

Chapter 1 WHOLE NUMBERS 1

Practice

10. (a) $9^2 = 81$ (b) $8^3 = 512$
   (c) $3^5 = 243$ (d) $3^3 \times 4^2 = 432$
   (e) $5^2 \times 2^4 = 400$

11. (a) 64 (b) 128
   (c) 400 (d) 216

12. Leo thought 6^3 meant 3 sixes. The sum of 3 sixes, that is, 3 groups of six, is 18. But 6^3 means the product of 6 multiplied by itself 3 times:
$6 \times 6 \times 6 = 216$

13. $810 \div 10 = 81$
$9^2 = 81$
The original number was 9.

Challenge

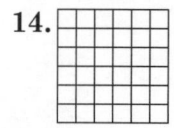

14.

15. (a) 2 (b) 5
   (c) 3

1.1C Order of Operations without Parentheses

Basics

16. (a) is the correct equation. The Order of Operations says we multiply before we add. The second equation was solved without considering the Order of Operations.

17. (a) $18 - 4 = 14$ (b) $20 - 5 = 15$
   (c) $1 \times 1 + 1 - 1$
   $= 1 + 1 - 1$
   $= 1$

Practice

18. (a) $14 + 12 \times 36$ or $12 \times 36 + 14$
   (b) $12 + 14 \times 36$ or $14 \times 36 + 12$
   (c) First order: $14 + 12 \times 36$
   $= 14 + 432 = 446$
   Second order: $12 + 14 \times 36$
   $= 12 + 504 = 516$
   The second order is for 70 more prizes than the first order.

19. (a) Third (b) First
   (c) Second

20. (a) $16 - 7 \times 2 + 8$
   $= 16 - 14 + 8$
   $= 10$
   (b) $50 + 81 \times 6 - 8$
   $= 50 + 486 - 8$
   $= 528$
   (c) $125 - 8 \times 7$
   $= 125 - 56$
   $= 69$
   (d) $23 + 25 \div 5 \times 32 - 7$
   $= 23 + 5 \times 32 - 7$
   $= 23 + 160 - 7$
   $= 176$
   (e) $4 + 9 - 3 \times 2$
   $= 4 + 9 - 6$
   $= 7$

Challenge

21. (a) $7 \times 12 + 3 \times 12 - 4 - 2$
$= 84 + 36 - 6$
$= 114$

(b) Answers will vary. Example:
$7 \times 12 + 3 \times 12 - 4 - 2$
$= 84 + 3 \times 12 - 4 - 2$
$= 87 \times 12 - 4 - 2$
$= 1{,}044 - 4 - 2$
$= 1{,}040 - 2$
$= 1{,}038$

22. (a) $10 \times 2 - 8 = 12$

(b) $(54 - 5) \div 7 = 7$

(c) First, think of a number that, when multiplied by itself is 81. That number is 9.
$9 + 11 = 20$

(d) First, think of a number that, when multiplied by itself three times is 64. That number is 4. Then think of a number that, when multiplied by itself is 4. That number is 2.
$64 \div 4 \div 4 \div 2 = 2$

1.1D Order of Operations with Parentheses

Basics

23. (a) $48 \div (3 + 5)$
(b) $(5 \times 9) \div (5 \times 3)$
(c) $2 \times 3 \times (15 - 9)$

24. $2^2 + 2 \times (3 + 3^2) \div 3$ matches to 12
$(2^2 + 2) \times (3 + 3^2) \div 3$ matches to 24
$2^2 + (2 \times 3) + 3^2 \div 3$ matches to 13

25. (a) 3×8
$= 24$
(b) $49 - 3 \times 3$
$= 49 - 9$
$= 40$
(c) $(4^2 + 2 \times 9) \div 2$
$= (16 + 18) \div 2$
$= 34 \div 2$
$= 17$

26. (a) $6 \times (3 + 3^3) = 180$
(b) $1 = (1 + 39) \div (16 - 6) \div 2^2$
(c) $(5 + 7) \times (8 - 6) \times 2 = 48$

Practice

27. (a) Answers for work shown may vary.
Example answer:
$50 - (25 + 4 + 16)$
$= 50 - 45 = 5$
Betty received $5 change.
(b) $7 \times 15 + 11 = 116$
There were 116 chairs altogether.

28. Cora solved by doing the multiplication and division first, and then the addition and subtraction.
$6 + 3 \times 6 \div 2 + 4 = 6 + 18 \div 2 + 4$
$= 6 + 9 + 4 = 19$
Cora was correct. Alyssa worked from left to right, ignoring the Order of Operations convention.

29. $1 + 2 + 3 + 4 + 5 = 15$
$1 + 2 + 3 + 4 + 5 + 6 = 21$
The number of dots in each of the next two patterns are 15 and 21.

30.

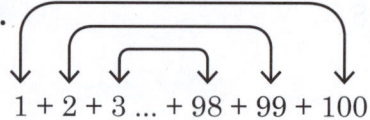

$1 + 2 + 3 \ldots + 98 + 99 + 100$

Each set of numbers sums to 101 and there are 50 sets.

$101 \times 50 = 5{,}050$

Answer: 5,050

Factors and Multiples

1.2A Factors

Basics

1. (a) 1, 2, 7, 14
    (b) 1, 2, 4, 8, 16, 32
    (c) 1, 2, 3, 4, 6, 8, 12, 16, 24, 32, 48, 96
    (d) 1, 3, 5, 7, 15, 21, 35, 105

2. (a) 9   (b) 14
    (c) 18

3. (a) 1, 2   (b) 1, 7

Practice

4. Using every flower, Taylor can make 7 arrangements with the same number of colored flowers in each arrangement. Each arrangement will have 3 yellow flowers, 5 pink flowers, and 6 white flowers.

5. Each strip of fabric should be 12 m long.

6. 1, 2, and 4 are common factors of 32, 40, and 60.
   The greatest number of bags Fang can make is 4.

Challenge

7. Answers may vary.
   Example answer: The ones digits increase by 1, the tens digits increase by 1 and the hundreds digits stay the same.
   For each product, the sum of the digits in the hundreds place and the ones place equals the digit in the tens place.

8. 12

1.2B Multiples

Basics

9. (a) 7, 14, 21, 28, 35
    (b) 12, 24, 36, 48, 60
    (c) 24, 48, 72, 96, 120

10. 3 {3, 6, 9, ⑫, 15, 18, 21, 24, 27, 30}
    4 {4, 8, ⑫, 16, 20, 24, 28, 32, 36, 40}

11. (a) 36   (b) 180

12. 60

Practice

13. Answers will vary.
    Example answers: He is right with 3 and 4. The least common multiple is 12. He is wrong with 4 and 6. The least common multiple is 12, not 24.

14. The least common multiple of 6 and 8 is 24. They will share the gym again in 24 days.

15. The least common multiple of 8 and 12 is 24. The first caller to win both will be the 24th caller.

Challenge

16. The least common multiple of 2, 7, and 12 is 84. The shortest possible length of each ribbon at the start is 84 in.

17. The least common multiple of 3, 4, and 6 is 12. They will start wrapping a gift at the same time again at 9:12 a.m.

18. The least common multiple of 3, 4, and 5 is 60. It will take 60 minutes until they all meet at the station.

19. The least common multiple of 9, 12, and 15 is 180.
    180 − 13 = 167
    The number that is less than the least common multiple of 9, 12, and 15 is 167.

Multiplication

1.3A Properties of Multiplication

Basics

1. (a) Distributive Property of Multiplication
   (b) Associative Property of Multiplication
   (c) Identity Property of Multiplication
   (d) Commutative Property of Multiplication
   (e) Zero Property of Multiplication

Practice

2. Solution methods may vary. An example solution method is shown for each problem.
   (a) (8 × 7) × (5 × 2) = 56 × 10
       = 560
   (b) 8 × 5 × 9 = 40 × 9
       = 360
   (c) (7 × 9) × (5 × 2) = 63 × 10
       = 630
   (d) (12 × 7) × (5 × 4 × 100) = 84 × 2,000
       = 168,000

3. (a) 7, 7   (b) 9, 9
   (c) The Distributive Property of Multiplication

Challenge

4. Method 1
   5 m × 12 m + 5 m × 7 m
   = 60 m^2 + 35 m^2
   = 95 m^2
   Method 2
   5 m × (12 + 7) m
   = 5 m × 19 m
   = 95 m^2

5. (a) 40 × $2 + 3 × (10 × $2) + 5 × (2 − $0.50)
   (b) 40 × $2 + 3 × (10 × $2) + 5 × (2 − $0.50)
       = $80 + $60 + $7.50 = $147.50
       Jenna does not have enough money.

1.3B Mental Multiplication

Basics

6. Solution methods will vary. Example answers:
   (a) 6 × (20 + 4)
       = (20 × 6) + (4 × 6)
       = 120 + 24
       = 144
   (b) 8 × (60 + 5)
       = (60 × 8) + (5 × 8)
       = 480 + 40
       = 520

7. **(a)** 7 × (70 − 2)
   = (70 × 7) − (2 × 7)
   = 476
   or:
   68 × (10 − 3)
   = (68 × 10) − (68 × 3)
   = 476

   (b) 8 × (30 − 1)
   = (30 × 8) − (1 × 8)
   = 232
   or:
   29 × (10 − 2)
   = (29 × 10) − (29 × 2)
   = 232

8. Answers will vary.
   Example answer:
   6 × 1,000 × 21
   = 6 × 21 × 1,000
   = 126 × 1,000
   = 126,000

Practice

9. Solution methods will vary.
   Example answers:
   (a) (30 + 8) × 6
   = 6 × 30 + 6 × 8
   = 180 + 48
   = 228

   (b) (30 + 6) × 7
   = 30 × 7 + 6 × 7
   = 210 + 42
   = 252

   (c) 8 × (100 + 20 + 6)
   = 8 × 100 + 8 × 20 + 8 × 6
   = 800 + 160 + 48
   = 1,008

   (d) 7 × (40 + 9)
   = 7 × 40 + 7 × 9
   = 280 + 63
   = 343
   or:
   7 × (50 − 1)
   = 7 × 50 − 7 × 1
   = 350 − 7
   = 343

10. Answers may vary.
    Example answers:
    (a) 2 × 1,000 × 24
    = 2 × 24 × 1,000
    = 48 × 1,000
    = 48,000

    (b) (80 + 3) × 7
    = 80 × 7 + 3 × 7
    = 560 + 21
    = 581

Challenge

11. The following are six different strategies:
    (a) 16 × 7
    = (8 × 7) + (8 × 7)
    = 56 + 56
    = 112

    (b) 16 × 7
    = (10 × 7) + (6 × 7)
    = 70 + 42
    = 112

    (c) 16 × 7
    = (20 × 7) − (4 × 7)
    = 140 − 28
    = 112

    (d) 16 × 7
    = (16 × 10) − (16 × 3)
    = 160 − 48
    = 112

(e) 16 × 7
= (16 × 2) + (16 × 2) + (16 × 2) + 16
= 32 + 32 + 32 + 16
= 112

(f) 16 × 7
= (16 × 5) + (16 × 2)
= 80 + 32
= 112

Division

1.4A Properties of Division

Basics

1. Dividend, divisor, quotient

2. (a) 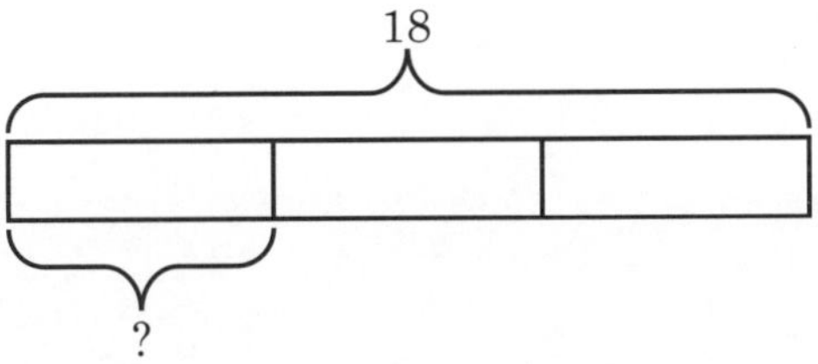
   18 yd ÷ 3 = 6 yd
   It took 6 yd of fabric to make each curtain.

   (b) 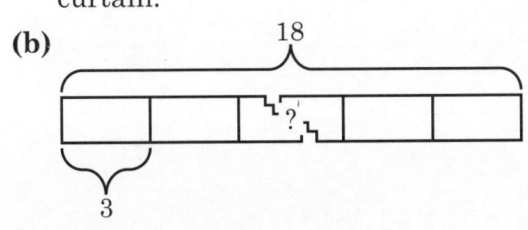
   18 yd ÷ 3 yd = 6
   Dan can make 6 curtains.

3. The quotient does not change. Examples will vary.
   One possible example:
   32 ÷ 8 = (32 ÷ 4) ÷ (8 ÷ 4) = 8 ÷ 2 = 4

4. (a) 7, 3   (b) 8, 6
   (c) 2, 5   (d) 4, 4

Practice

5. 56 ÷ 8 = 7
   Janet can use 7 bags.

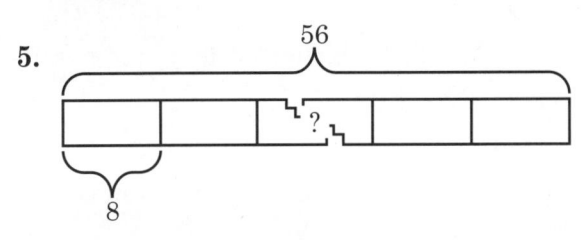

6. (72 in² + 40 in²) ÷ 14 in
   = 112 in² ÷ 14 in = 8 in
   The width of the rectangle is 8 in.

Challenge

7. Identity Property of Division: When we divide a number by 1, we get the same number. 100 ÷ 1 = 100

   When we divide a number by itself, we get 1. 100 ÷ 100 = 1

   Zero Property of Division: When we divide 0 by a non-zero number, we will get 0. 0 ÷ 100 = 0

1.4B Mental Division

Basics

8. (a) 15, 15, 15   (b) 50, 1

Practice

9. Answers may vary. Below is one method for each problem.
   (a) (80 ÷ 8) + (16 ÷ 8) = 10 + 2 = 12
   (b) (90 ÷ 3) + (12 ÷ 3) = 30 + 4 = 34
   (c) (210 ÷ 3) + (6 ÷ 3) = 70 + 2 = 72
   (d) (250 ÷ 5) + (35 ÷ 5) = 50 + 7 = 57
   (e) (75,000 ÷ 10) ÷ (250 ÷ 10)
       = 7,500 ÷ 25 = 300

10. Explanations may vary. Sample explanations:
    (a) True.
        Divide both the dividend and the divisor by 2,000.
    (b) True.
        Divide both the dividend and the divisor by 300.

11. Answers may vary. Below are two possibilities. Students may choose to divide the dividend and denominator by 3, 5, 9, 15, or 45.
    $225 \div 45 = (225 \div 5) \div (45 \div 5)$
    $= 45 \div 9 = 5$
    $225 \div 45 = (225 \div 9) + (45 \div 9)$
    $= 25 \div 5 = 5$

12. (a) 16   (b) 68
    (c) 15   (d) 70
    (e) 11,000

13. Both girls' methods ultimately divide 636 by 6. Mariam broke apart the dividend into 600 and 36, and divided each part by 6. Riya broke apart the divisor, 6, into two factors (2 and 3).

Challenge

14. Tim is correct. Explanations may vary. Sample explanation: 1,920 and 1,080 have a common factor of 120.
    $1,920 \div 120 = 16$ and $1,080 \div 120 = 9$.

15. (a) 63   (b) 90
    (c) 9,000

Chapter 2: Fractions
Multiplication of Fractions

2.1A Multiplication of a Proper Fraction by a Whole Number

Basics

1.

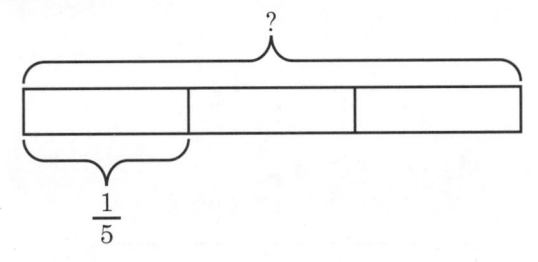

$\frac{1}{5}$

Method 1
1 unit = $\frac{1}{5}$

3 units = $3 \times \frac{1}{5} = \frac{3}{5}$

Method 2
$3 \times \frac{1}{5} = \frac{3 \times 1}{5} = \frac{3}{5}$

2.

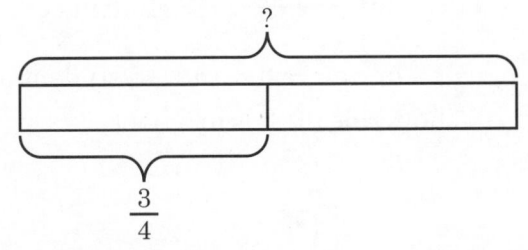

$\frac{3}{4}$

Method 1
From the model,
1 unit = $\frac{3}{4}$
2 units = $2 \times \frac{3}{4} = \frac{6}{4} = 1\frac{1}{2}$

Method 2
$\frac{3}{4} \times 2 = \frac{3 \times 2}{4} = \frac{6}{4} = 1\frac{1}{2}$

3.

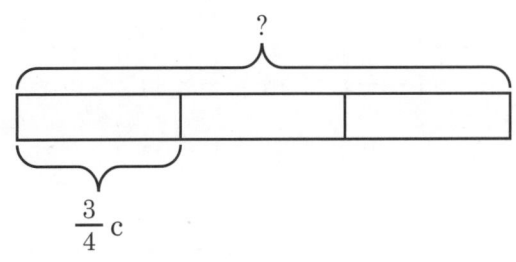

$\frac{3}{4}$ c

From the model,
1 unit = $\frac{3}{4}$ c

3 units = $\frac{3}{4}$ c $\times 3 = \frac{3 \times 3}{4}$ c $= \frac{9}{4}$ c $= 2\frac{1}{4}$ c

May used $2\frac{1}{4}$ c of raisins.

4. $\frac{2}{5}$ m

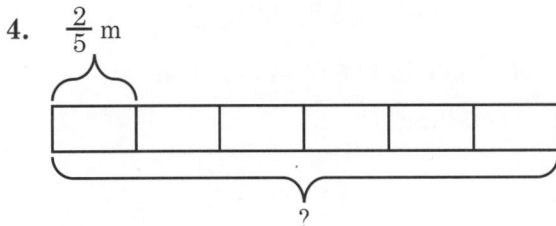

From the model,
1 unit = $\frac{2}{5}$ m

6 units = $\frac{2}{5}$ m $\times 6 = \frac{2 \times 6}{5}$ m $= \frac{12}{5}$ m
= $2\frac{2}{5}$ m

Hector needs $2\frac{2}{5}$ m of wood.

Practice

5. $48 \times \frac{2}{3}$ mi = $\frac{48 \times 2}{3}$ mi = 16×2 mi = 32 mi

Julia walked 32 miles.

6. **Method 1**
   $(1 - \frac{2}{3}) \times 345 = \frac{1}{3} \times 345 = 115$

   Method 2

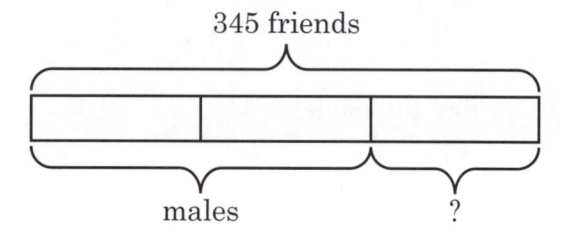

   345 friends

   males ?

   From the model,
   3 units = 345
   1 unit = $\frac{345}{3}$ = 115

   Caleb has 115 female friends on social media.

7. 522 cars

   New ?

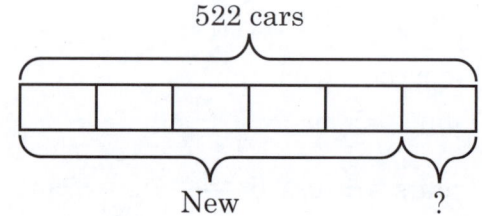

   Method 1
   From the model,
   6 units = 522 cars
   1 unit = $\frac{522}{6}$ = 87 cars

   Method 2
   $\frac{1}{6} \times 522 = \frac{522}{6} = 87$ cars

   87 of the cars are used.

Challenge

8. **Method 1**

   $1 - \frac{5}{8} = \frac{3}{8}$

   $\frac{3}{8} \times \cancel{240}^{30} = 90$ boys

   There are 90 boys.

   $\frac{1}{\cancel{8}_1} \times \cancel{90}^{45} = 45$ boys

   Method 2

   240 students

   Girls   Boys riding bus

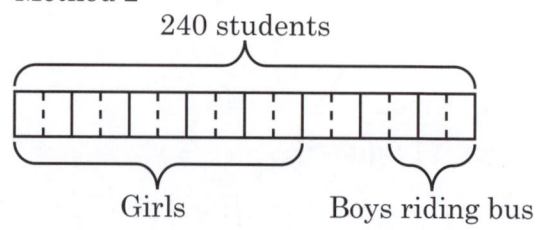

   From the model,
   16 units → 240 students
   1 unit → $\frac{240}{16}$ = 15 students
   3 units → 3 × 15 = 45 students

   45 boys ride the bus to school.

9. **Method 1**

   The amount of money spent on gift:
   $\frac{5}{8} \times \$424 = \$265$

   Remaining amount of money:
   $\$424 - \$265 = \$159$

   Amount of money spent on the book:
   $\frac{1}{3} \times \$159 = \$53$

   $\$265 - \$53 = \$212$

   Method 2

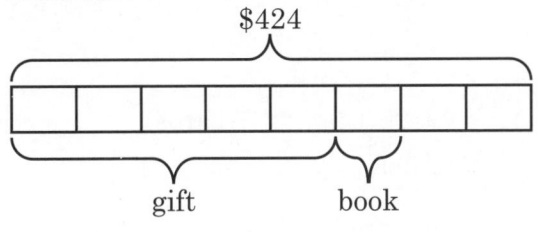

   $424 is divided into 8 units. The gift is 5 units and the book is 1 unit of the remaining 3 units.

   The gift cost 4 units more than the book.

   From the model,
   8 units → $424

   1 unit → $\frac{\$424}{8} = \$53$

   4 units → $4 \times \$53 = \$212$

   The gift cost $212 more than the book.

2.1B Multiplication of a Proper Fraction by a Fraction

Basics

10. (a) $\frac{1 \times 1}{4 \times 3} = \frac{1}{12}$ (b) $\frac{1 \times 2}{5 \times 3} = \frac{2}{15}$

    (c) $\frac{\cancel{4}^1 \times \cancel{3}^1}{\cancel{5}_1 \times \cancel{8}_2} = \frac{1}{2}$ (d) $\frac{\cancel{6}^2 \times 2}{7 \times \cancel{8}_1} = \frac{4}{7}$

Practice

11. $\frac{1 \times 3}{2 \times 5} = \frac{3}{10}$

    $\frac{3}{10}$ of all of the vehicles for sale are white trucks.

12. $\frac{1 \times \cancel{4}^1}{\cancel{4}_1 \times 7} = \frac{1}{7}$

    $\frac{1}{7}$ of the necklaces are bead necklaces.

    $\frac{1 \times 1}{7 \times 7} = \frac{1}{49}$

    $\frac{1}{49}$ of her jewelry pieces are necklaces made of green beads.

Challenge

13. $\frac{\cancel{8}^1 \times \cancel{8}^2}{\cancel{4}_1 \times \cancel{9}_3} \text{ m} = \frac{2}{3} \text{ m}$

    $\frac{2}{3}$ m is painted red.

    $\frac{8}{9} \text{ m} - \frac{2}{3} \text{ m} = \frac{8-6}{9} \text{ m} = \frac{2}{9} \text{ m}$

    $\frac{\cancel{8}^1 \times \cancel{2}^1}{\cancel{8}_4 \times \cancel{9}_3} \text{ m} = \frac{1}{12} \text{ m}$

    $\frac{1}{12}$ m of the rod is not painted.

14. $\frac{1}{4} \times \frac{1}{2} = \frac{1}{8}$

    $\frac{1}{3} \times \frac{1}{8} = \frac{1}{24}$

    $1 - \frac{1}{24} = \frac{23}{24}$

    $\frac{23}{24}$ of the entire garden is not used to plant red roses.

2.1C Multiplication of an Improper Fraction or a Mixed Number by a Whole Number

Basics

15. (a) $\frac{7}{5} \times 3 = \frac{7 \times 3}{5} = \frac{21}{5} = 4\frac{1}{5}$

(b) $\frac{31}{\cancel{8}_2} \times \cancel{4}^1 = \frac{31}{2} = 15\frac{1}{2}$

(c) $\cancel{6}^1 \times \frac{17}{\cancel{12}_2} = \frac{17}{2} = 8\frac{1}{2}$

(d) $\cancel{15}^3 \times \frac{21}{\cancel{10}_2} = \frac{3 \times 21}{10} = \frac{63}{2} = 31\frac{1}{2}$

Practice

16. $4 \times 2\frac{2}{3}$ kg $= 4 \times \frac{8}{3}$ kg $= \frac{32}{3}$ kg $= 10\frac{2}{3}$ kg

The combined weight of the packages that Wyatt wants to mail is $10\frac{2}{3}$ kg.

17. $12 \times 2\frac{3}{8}$ oz

$= \cancel{12}^3 \times \frac{19}{\cancel{8}_2}$ oz $= \frac{3 \times 9}{2}$ oz $= \frac{57}{2}$ oz $= 28\frac{1}{2}$ oz

Linda bought $28\frac{1}{2}$ oz of seeds.

18. Method 1

$3 \times 1\frac{1}{4}$ lb $= 3 \times \frac{5}{4}$ lb $= \frac{3 \times 5}{4}$ lb $= \frac{15}{4}$ lb

$\cancel{2}^1 \times \frac{15}{\cancel{4}_2}$ lb $= \frac{15}{2}$ lb $= 7\frac{1}{2}$ lb

Method 2

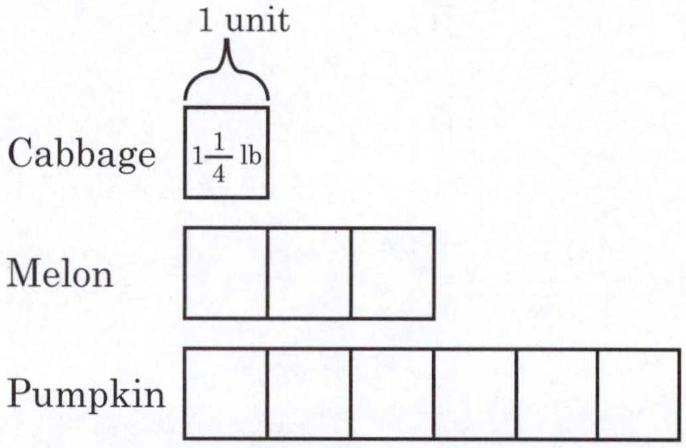

Cabbage: 1 unit, $1\frac{1}{4}$ lb

Melon: 2 units

Pumpkin: 6 units

From the model,
1 unit = $1\frac{1}{4}$ lb
6 units → $6 \times 1\frac{1}{4}$ lb = $7\frac{1}{2}$ lb

The pumpkin weighs $7\frac{1}{2}$ lb.

Challenge

19. Method 1

Baseball cards:

$\frac{5}{3} \times 45 = \frac{5 \times 45}{3} = \frac{225}{3} = 75$ cards

Football cards and baseball cards:

$45 + 75 = 120$ cards

Hockey cards:

$\frac{3}{4} \times 120 = \frac{3 \times 120}{3} = \frac{360}{4} = 90$ cards

Total cards:

$120 + 90 = 210$ cards

Method 2

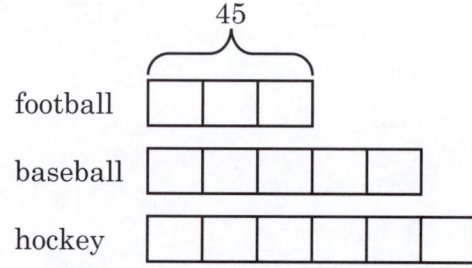

football: 3 units (45)
baseball: 5 units
hockey: 6 units

$\frac{3}{4}$ as many hockey cards means $\frac{3}{4}$ of 8 units, which is 6 units.

From the model,
3 units = 45 cards

1 unit → $\frac{45}{3} = 15$ cards

14 units → $14 \times 15 = 210$ cards

There are 210 cards in Josef's collection.

20. $\frac{3}{7}$ as many nonfiction as fiction means for 4 units of fiction, there are 7 units of nonfiction.

Method 1

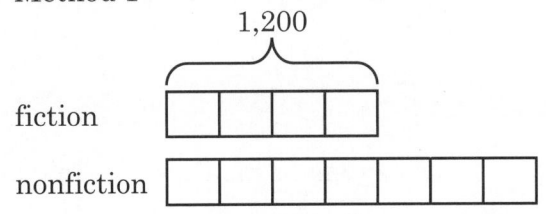

```
            1,200
         ⌐‾‾‾‾‾‾‾⌐
fiction    [ ][ ][ ][ ]
nonfiction [ ][ ][ ][ ][ ][ ][ ]
```

From the model,
4 units = 1,200 books
1 unit → $\frac{1,200}{4}$ = 300 books
7 units → 7 × 300 = 2,100 books
There are 2,100 nonfiction books.

$\frac{1}{\cancel{3}_1} \times \cancel{2,100}^{700}$ = 700 books

Method 2

$1\frac{3}{4} \times 1,200 = 2,100$ books

$\frac{1}{\cancel{3}_1} \times \cancel{2,100}^{700}$ = 700 books

There are 700 history books.

2.1D Multiplication of an Improper Fraction or a Mixed Number by a Fraction

Basics

21. (a) $\frac{29}{\cancel{6}_2} \times \frac{\cancel{3}^1}{4} = \frac{29 \times 1}{2 \times 4} = \frac{29}{8}$
    $= 3\frac{5}{8}$

(b) $\frac{15}{8} \times \frac{3}{4} = \frac{15 \times 3}{8 \times 4} = \frac{45}{32}$
    $= 1\frac{13}{32}$

(c) $\frac{17}{6} \times \frac{1}{3}$
    $= \frac{17}{18}$

(d) $\frac{\cancel{3}^1}{\cancel{4}_1} \times \frac{\cancel{16}^4}{\cancel{9}_3}$
    $= 1\frac{1}{3}$

22. (a) $\frac{\cancel{18}^2}{\cancel{5}_1} \times \frac{\cancel{35}^7}{\cancel{9}_1}$
    $= 14$

(b) $\frac{4}{\cancel{8}_1} \times \frac{\cancel{15}^5}{7} = \frac{20}{7}$
    $= 2\frac{6}{7}$

(c) $\frac{73}{\cancel{9}_1} \times \frac{\cancel{9}^1}{5} = \frac{73}{5}$
    $= 14\frac{3}{5}$

Practice

23. $1 - \frac{1}{4} = \frac{3}{4}$

$\frac{3}{4} \times 3\frac{2}{3}$ tons $= \frac{\cancel{3}^1}{4} \times \frac{11}{\cancel{3}_1}$ tons $= \frac{11}{4}$ tons
$= 2\frac{3}{4}$ tons

Mr. King used $2\frac{3}{4}$ tons of gravel in his side yard.

24. $6\frac{5}{8}$ ft $\times 2\frac{2}{3}$ ft $= \frac{53}{\cancel{8}_1}$ ft $\times \frac{\cancel{8}^1}{3}$ ft $= 17\frac{2}{3}$ ft²

The area of the field is $17\frac{2}{3}$ ft².

Challenge

25. The difference between what Kamala had left and what she gave to her brother is $\frac{2}{5}$ L.

$$4\frac{4}{5} \text{ L} \times \frac{1}{4} = \frac{\cancel{24}^{6}}{5} \text{ L} \times \frac{1}{\cancel{4}_{1}} = \frac{6}{5} \text{ L} = 1\frac{1}{5} \text{ L}$$

Kamala's brother's portion was $1\frac{1}{5}$ L.

$$4\frac{4}{5} \text{ L} - 1\frac{1}{5} \text{ L} = 3\frac{3}{5} \text{ L}$$

$$\frac{1}{9} \times 3\frac{3}{5} \text{ L} = \frac{1}{\cancel{9}_{1}} \times \frac{\cancel{18}^{2}}{5} \text{ L} = \frac{2}{5} \text{ L}$$

Kamala drank $\frac{2}{5}$ L.

$$3\frac{3}{5} \text{ L} - \frac{2}{5} \text{ L} = 3\frac{1}{5} \text{ L}$$

$$\frac{1}{2} \times 3\frac{1}{5} \text{ L} = \frac{1}{\cancel{2}_{1}} \times \frac{\cancel{16}^{8}}{5} \text{ L} = \frac{8}{5} \text{ L} = 1\frac{3}{5} \text{ L}$$

Kamala had $1\frac{3}{5}$ L left.

$$1\frac{3}{5} \text{ L} - 1\frac{1}{5} \text{ L} = \frac{2}{5} \text{ L}$$

The difference is $\frac{2}{5}$ L.

Division of Fractions

2.2A Division of a Whole Number by a Fraction

Basics

1. (a) Method 1

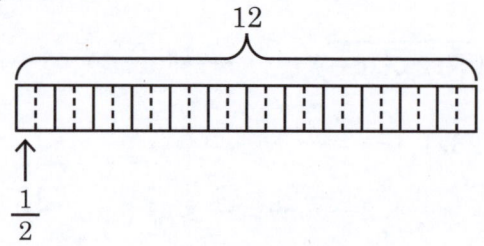

From the model, we see that there are 24 $\frac{1}{2}$s in 12.

$$12 \div \frac{1}{2} = 24$$

Method 2

$$12 \div \frac{1}{2} = 12 \times \frac{2}{1} = 24$$

24

(b) Method 1

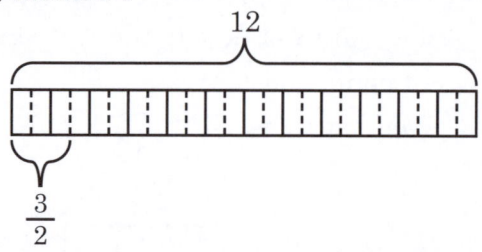

From the model, we see that there are 8 $\frac{3}{2}$s in 12.

$$12 \div \frac{3}{2} = 8$$

Method 2

$$12 \div \frac{3}{2} = 12 \times \frac{2}{3} = 8$$

8

(c) Method 1

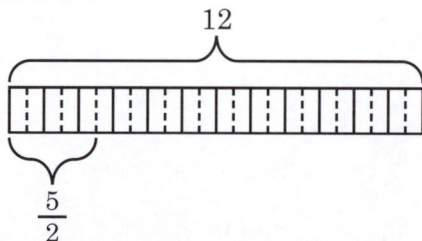

From the model, we see that there are $4\frac{4}{5}$ $\frac{5}{2}$s in 12.

$$12 \div \frac{5}{2} = 8$$

Method 2

$$12 \div \frac{5}{2} = 12 \times \frac{2}{5} = 4\frac{4}{5}$$

$4\frac{4}{5}$

2. **(a)** Method 1

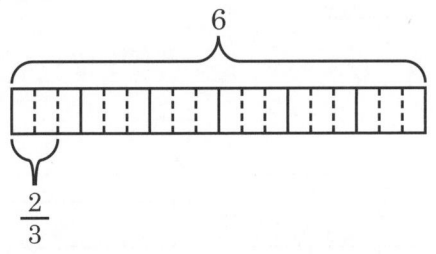

From the model, we see that there are 9 $\frac{2}{3}$s in 6.

Method 2

$6 \div \frac{2}{3} = 6 \times \frac{3}{2} = 9$

9

(b) Method 1

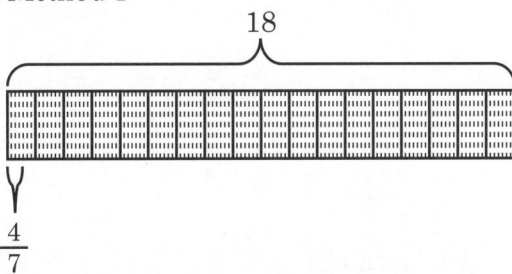

From the model, we see that there are $31\frac{1}{2}$ $\frac{4}{7}$s in 18.

Method 2

$18 \div \frac{4}{7} = \cancel{18}^9 \times \frac{7}{\cancel{4}_2} = 9 \times \frac{7}{2} = \frac{63}{2}$

$= 31\frac{1}{2}$

$31\frac{1}{2}$

(c) Method 1

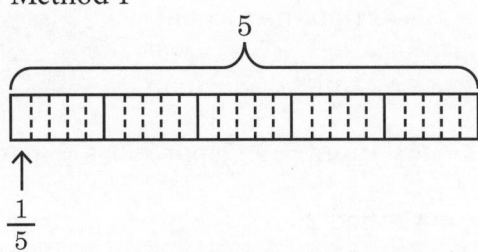

From the model, we see that there are 25 $\frac{1}{5}$s in 5.

Method 2

$5 \div \frac{1}{5} = 5 \times \frac{5}{1} = 25$

25

Practice

3. Method 1

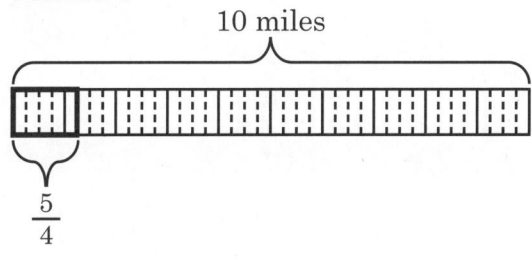

From the model,

$\frac{5}{4}$ mile ⟶ 1 stop

10 miles ⟶ 8 stops

Method 2

$10 \div \frac{5}{4} = 10 \times \frac{4}{5} = \frac{10 \times 4}{5} = \frac{40}{5} = 8$

Emily made 8 stops.

4. $27 \div \frac{3}{5} = 27 \times \frac{5}{3} = 45$ bags

There were 45 bags.

5. Method 1

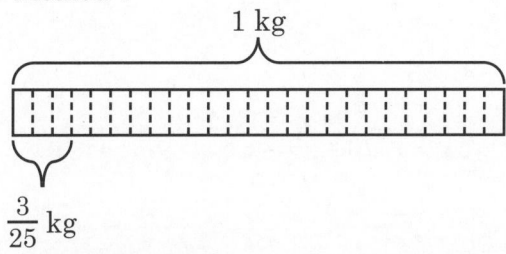

From the model,

$\frac{3}{25}$ kg ⟶ 1 c

$\frac{25}{3} = 8\frac{1}{3}$ c

Method 2

$1 \div \frac{3}{25} = 1 \times \frac{25}{3}$ c $= \frac{25}{3}$ c $= 8\frac{1}{3}$ c

The baker has $8\frac{1}{3}$ cups of flour.

Chapter 2 FRACTIONS   15

Challenge

6. Method 1

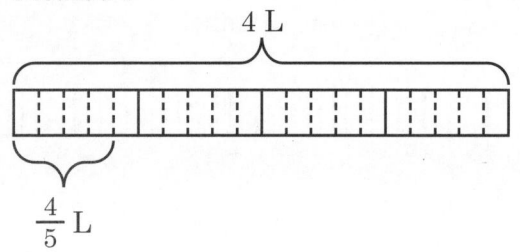

[Bar model: 4 L total divided into 5 equal parts, each $\frac{4}{5}$ L]

From the model,

$\frac{4}{5}$ L ⟶ 1 container

4 L ⟶ 5 containers

Method 2

$4 \div \frac{4}{5} = 4 \times \frac{5}{4} = 5$ containers

Ms. Aquino will fill 5 small containers.

2.2B Division of a Fraction by a Whole Number

Basics

7. $\frac{2}{3} \div 5$

8. (a) $\frac{4}{5} \div 20 = \frac{\cancel{4}^{1}}{5} \times \frac{1}{\cancel{20}_{5}}$

$= \frac{1}{25}$

(b) $\frac{5}{4} \times \frac{1}{3}$

$= \frac{5}{12}$

(c) $\frac{1}{5} \times \frac{1}{5}$

$= \frac{1}{25}$

Practice

9. Method 1

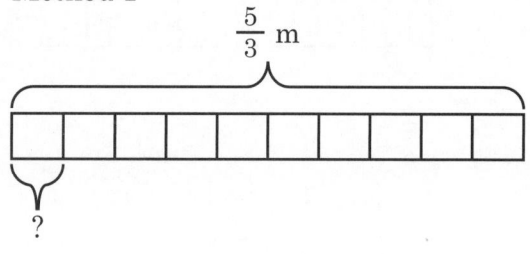

[Bar model: total $\frac{5}{3}$ m divided into 10 units, 1 unit labeled ?]

From the model,

10 units = $\frac{5}{3}$ m

1 unit ⟶ $\frac{5}{3}$ m ÷ 10 = $\frac{1}{6}$ m

Method 2

$\frac{5}{3}$ m ÷ 10 = $\frac{\cancel{5}^{1}}{3}$ m × $\frac{1}{\cancel{10}_{2}} = \frac{1}{6}$ m

The bush grew $\frac{1}{6}$ m each year.

10. Method 1

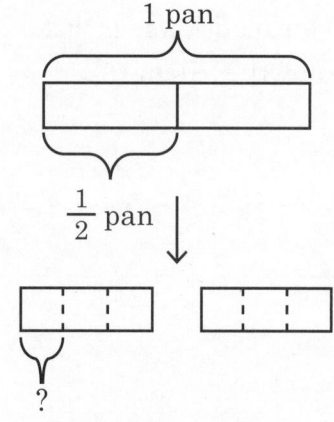

[Bar model: 1 pan divided in half, $\frac{1}{2}$ pan then divided into 3 units, 1 unit labeled ?]

From the model,

3 units ⟶ $\frac{1}{2}$ pan

1 unit ⟶ $\frac{1}{2}$ pan ÷ 3 = $\frac{1}{6}$ pan

Method 2

$\frac{1}{2} \div 3 = \frac{1}{2} \times \frac{1}{3} = \frac{1}{6}$ pan

Each friend got $\frac{1}{6}$ of the pan of brownies.

11. Method 1

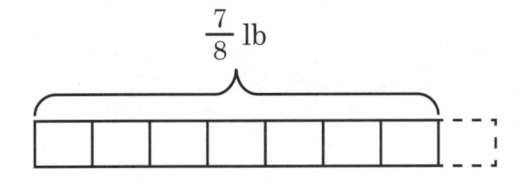

$\frac{7}{8}$ lb

From the model,

7 units → $\frac{7}{8}$ lb

1 unit → $\frac{7}{8}$ lb ÷ 7 = $\frac{1}{8}$ lb

Method 2

$\frac{7}{8}$ lb ÷ 7 = $\frac{7}{8}$ lb × $\frac{1}{7}$ = $\frac{1}{8}$ lb

$\frac{1}{8}$ lb of rice is in each bag.

12. $\frac{7}{9}$ yd ÷ 9 = $\frac{7}{9}$ yd × $\frac{1}{9}$ = $\frac{7}{81}$ yd

Each piece is $\frac{7}{81}$ yd long.

Challenge

13.

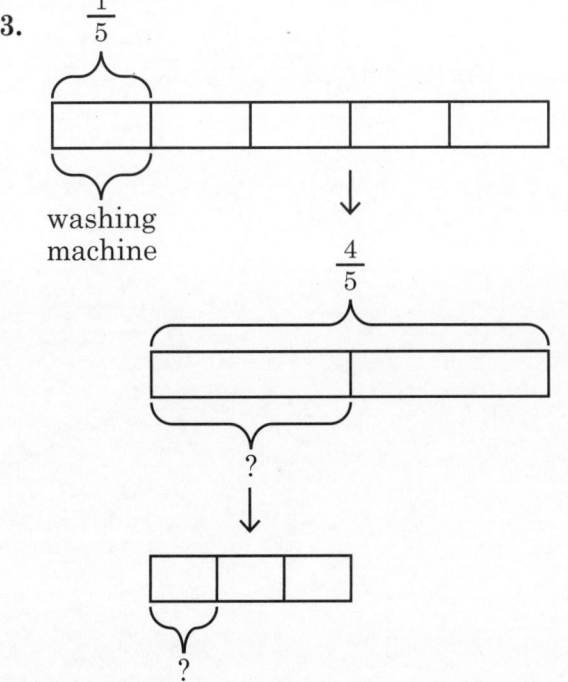

$\frac{4}{5}$ ÷ 2 = $\frac{4}{5}$ × $\frac{1}{2}$ = $\frac{2}{5}$

$\frac{2}{5}$ ÷ 3 = $\frac{2}{5}$ × $\frac{1}{3}$ = $\frac{2}{15}$

Each child received $\frac{2}{15}$ of her savings.

2.2C Division of a Fraction by a Fraction

Basics

14. (a)

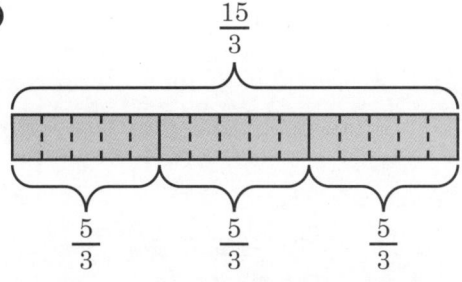

From the model, there are 3 groups of $\frac{5}{3}$.
Therefore, $\frac{15}{3}$ ÷ $\frac{5}{3}$ = 3.

(b)

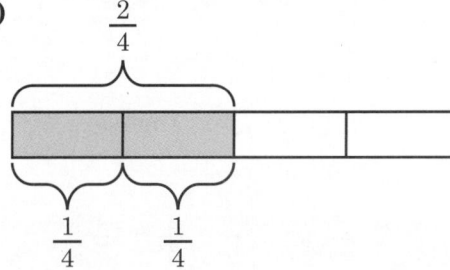

From the model, there are 2 groups of $\frac{1}{4}$.
Therefore, $\frac{2}{4}$ ÷ $\frac{1}{4}$ = 2.

(c)

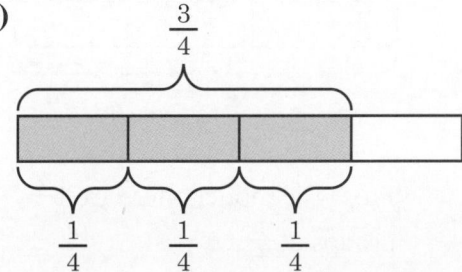

From the model, there are 3 groups of $\frac{1}{4}$.
Therefore, $\frac{3}{4}$ ÷ $\frac{1}{4}$ = 3.

(d)

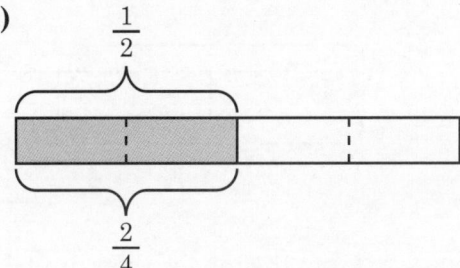

From the model, there is 1 group of $\frac{2}{4}$.
Therefore, $\frac{2}{4}$ ÷ $\frac{1}{2}$ = 1

Chapter 2 FRACTIONS

15. (a)

From the model, there are 7 groups of $\frac{1}{6}$.

Therefore, $\frac{7}{6} \div \frac{1}{6} = 7$.

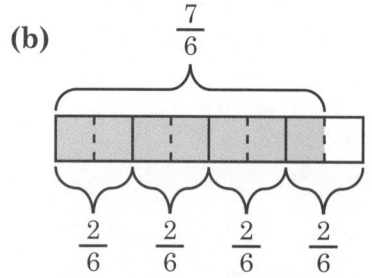

(b)

From the model, there are $3\frac{1}{2}$ groups of $\frac{2}{6}$.

Therefore, $\frac{7}{6} \div \frac{2}{6} = 3\frac{1}{2}$.

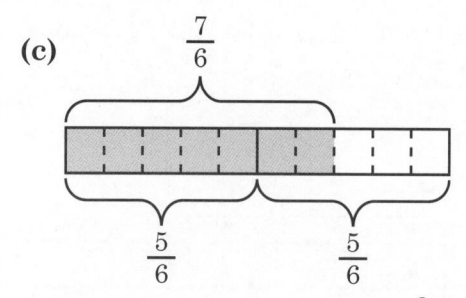

(c)

From the model, there are $1\frac{2}{5}$ groups of $\frac{5}{6}$.

Therefore, $\frac{7}{6} \div \frac{5}{6} = 1\frac{2}{5}$.

(d)

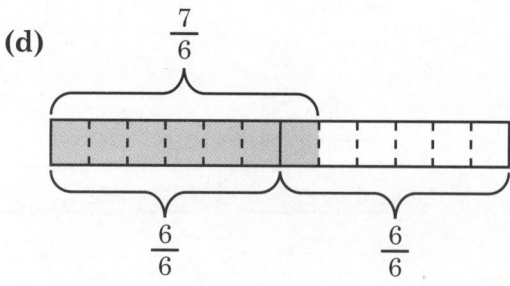

From the model, there are $1\frac{1}{6}$ groups of $\frac{6}{6}$.

Therefore, $\frac{7}{6} \div \frac{6}{6} = 1\frac{1}{6}$.

16. (a) $\frac{2}{6} \div \frac{1}{6} = \frac{2}{6} \times \frac{6}{1} = \frac{2}{1} \times \frac{1}{1}$
$= 2$

(b) $\frac{10}{8} \div \frac{5}{8} = \frac{10}{8} \times \frac{8}{5} = \frac{2}{1} \times \frac{1}{1}$
$= 2$

(c) $\frac{9}{10} \div \frac{2}{10} = \frac{9}{10} \times \frac{10}{2} = \frac{9}{1} \times \frac{1}{2} = \frac{9}{2}$
$= 4\frac{1}{2}$

(d) $\frac{11}{7} \div \frac{4}{7} = \frac{11}{7} \times \frac{7}{4} = \frac{11}{1} \times \frac{1}{4} = \frac{11}{4}$
$= 2\frac{3}{4}$

17. (a) $\frac{3}{7} \div \frac{1}{2} = \frac{3}{7} \times \frac{2}{1}$
$= \frac{6}{7}$

(b) $\frac{5}{8} \div \frac{7}{6} = \frac{5}{8} \times \frac{6}{7} = \frac{5}{4} \times \frac{3}{7}$
$= \frac{15}{28}$

(c) $1\frac{1}{8} \div \frac{1}{4} = \frac{9}{8} \times \frac{4}{1} = \frac{9}{2} \times \frac{1}{1} = \frac{9}{2}$
$= 4\frac{1}{2}$

(d) $1\frac{2}{3} \div 1\frac{3}{5} = \frac{5}{3} \times \frac{5}{8} = \frac{25}{24}$
$= 1\frac{1}{24}$

Practice

18. Quotient = $1\frac{1}{2}$

Word problems and models will vary.

Example problem:

Marlena has walked $\frac{3}{4}$ mi, which is $\frac{1}{2}$ of the total hike. What is the distance of the total hike?

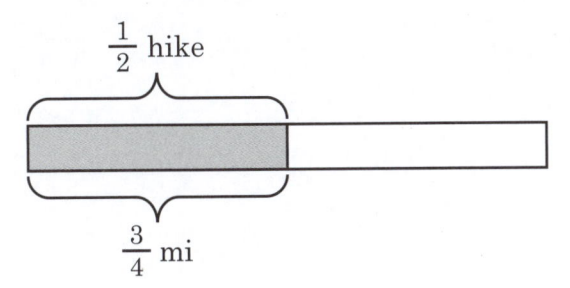

From the model,

$\frac{1}{2}$ hike ⟶ $\frac{3}{4}$ mi
Total hike ⟶ $\frac{3}{4}$ mi × 2 = $1\frac{1}{2}$ mi

The total hike is $1\frac{1}{2}$ mi.

19.

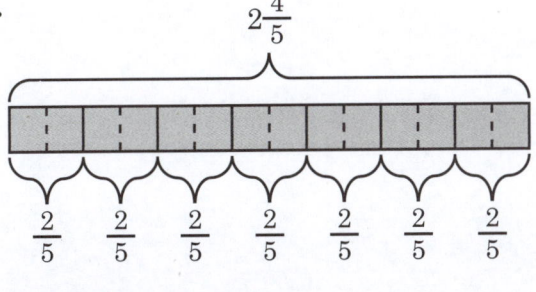

$2\frac{4}{5} \div \frac{2}{5} = \frac{14}{5} \times \frac{5}{2} = 7$

Pablo will end up with 7 pieces of rope.

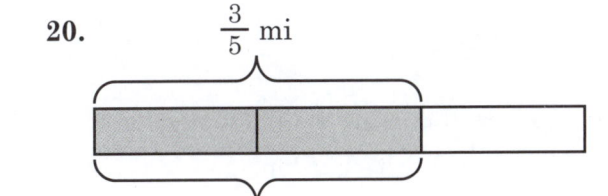

20.

From the model,

$\frac{2}{3}$ hr ⟶ $\frac{3}{5}$ mi

2 units ⟶ $\frac{3}{5}$ mi

$\frac{1}{3}$ hr ⟶ $\frac{3}{5}$ mi ÷ 2 = $\frac{3}{10}$ mi

1 unit ⟶ $\frac{3}{5}$ mi ÷ 2 = $\frac{3}{5}$ mi × $\frac{1}{2}$
= $\frac{3}{10}$ mi

1 hr ⟶ $\frac{3}{10}$ mi × 3 = $\frac{9}{10}$ mi

3 units ⟶ $\frac{3}{10}$ mi × 3 = $\frac{9}{10}$ mi

At that rate, a turtle would crawl $\frac{9}{10}$ of a mile in an hour.

21.

$12\frac{1}{2}$ ft² ÷ $2\frac{3}{4}$ ft = $\frac{25}{2}$ ft² ÷ $\frac{11}{4}$ ft

= $\frac{25}{2}$ ft² × $\frac{4}{11 \text{ ft}}$

= $\frac{50}{11}$ ft

= $4\frac{6}{11}$ ft

The length of the picture is $4\frac{6}{11}$ ft.

Challenge

22. $29\frac{6}{25}$ ft of fencing would be needed to fence the entire garden.

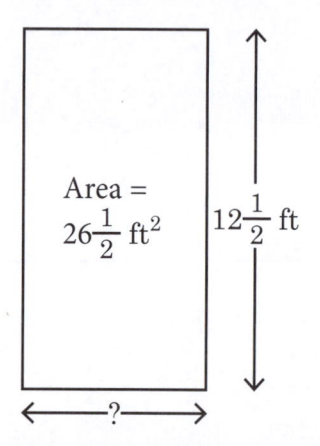

[Diagram: rectangle with Area = $26\frac{1}{2}$ ft², height $12\frac{1}{2}$ ft, width = ?]

$$26\frac{1}{2} \text{ ft}^2 \div 12\frac{1}{2} \text{ ft} = \frac{53}{2} \text{ ft}^2 \div \frac{25}{2} \text{ ft}$$

$$= \frac{53}{\underset{1}{\cancel{2}}} \text{ ft}^2 \times \frac{\cancel{2}^1}{25 \text{ ft}}$$

$$= \frac{53}{25} \text{ ft}$$

$$= 2\frac{3}{25} \text{ ft}$$

The width of the garden is $2\frac{3}{25}$ ft.

$2\frac{3}{25}$ ft + $2\frac{3}{25}$ ft + $12\frac{1}{2}$ ft + $12\frac{1}{2}$ ft

$= 29\frac{6}{25}$ ft

$29\frac{6}{25}$ ft of fencing would be needed.

Chapter 3: Decimals
Addition and Subtraction of Decimals

3.1A Structure of Decimals

Basics

1. (a) 3, 5   (b) 0, 0, 1
   (c) 6, 0, 2   (d) 8, 7
   (e) 0, 4, 4

2. (a) 0.9   (b) 2.3
   (c) 0.45   (d) 5.2
   (e) 3.05   (f) 0.032

3. (a) $\frac{3 \times 25}{4 \times 25} = \frac{75}{100}$
   $= 0.75$
   (b) $\frac{44 \times 2}{50 \times 2} = \frac{88}{100}$
   $= 0.88$
   (c) $3 + \frac{1 \times 2}{5 \times 2} = 3 + \frac{2}{10}$
   $= 3.2$
   (d) $\frac{9 \times 125}{8 \times 125} = \frac{1,125}{1,000}$
   $= 1.125$
   (e) $\frac{15 \times 25}{4 \times 25} = \frac{375}{100}$
   $= 3.75$
   (f) $4 + \frac{15 \times 5}{20 \times 5} = 4 + \frac{75}{100}$
   $= 4.75$

4. (a) $6\frac{17}{100}$ (b) $3\frac{60}{100} = 3\frac{3}{5}$
   (c) $\frac{24}{100} = \frac{6}{25}$ (d) $13\frac{225}{1,000} = 13\frac{9}{40}$
   (e) $2\frac{45}{100} = 2\frac{9}{20}$ (f) $8\frac{28}{1,000} = 8\frac{7}{250}$

Practice

5. (a) 0.302   (b) 4.5
   (c) 20.372

6. (a) $0.001, \frac{1}{100}, 0.1, 1\frac{1}{10}$
   (b) $16.02, 16.022, 16\frac{5}{25}, 16\frac{22}{100}$

Challenge

7. Answers will vary.
   Example answers: 24.154, 12.382

3.1B Adding and Subtracting Decimals

Basics

8. (a) 21.305   (b) 77.771
   (c) 27.025   (d) 5.399

9. (a) 1.06 + 0.5 = 1.56
   (b) 32.5 − 30.25 = 2.25
   (c) 1.2 + 0.03 + 0.004 = 1.234
   (d) 0.25 − 0.025 = 0.225

Practice

10. $23.88 + $42.09 + $33.77 = $99.74
    $100 − $99.74 = $0.26
    The value of Grace's gift card will be $0.26.

11. 3.52 m + 17.3 m + 27.009 m = 47.829 m
    50 m − 47.829 m = 2.171 m
    Hannah does not have enough ribbon.
    She needs 2.171 meters more.

12. Longer trail: 6.45 mi + 2.08 mi = 8.53 mi
    Combined distance:
    8.53 mi + 6.45 mi = 14.98 mi
    The combined distance is 14.98 miles.

13. (a) 3; 0; 7; 4   (b) 3,074
    (c) 307; 4   (d) 30; 7; 4

14.

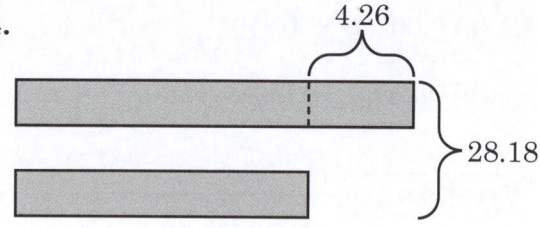

28.18 − 4.26 = 23.92

2 units ⟶ 23.92

1 unit ⟶ $\frac{23.92}{2}$ = 11.96

11.96 + 4.26 = 16.22

One number is 11.96 and the other is 16.22.

Challenge

15. 3.247 km + 3.9 km + 4.12 km + 3.006 km + (3.247 km + 0.9 km) = 18.42 km

20 km − 18.42 km = 1.58 km

He missed his goal by 1.58 km.

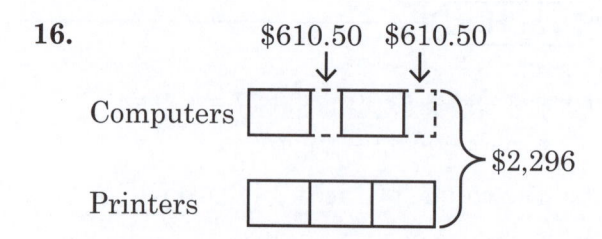

16.

From the model,

5 units = $2,296 − $610.50 − $610.50
       = $1,075

1 unit ⟶ $\frac{\$1,075}{5}$ = $215

3 units ⟶ 3 × $215 = $645

2 units ⟶ 2 × $215 = $430

The cost of all computers is:
$430 + $1,221 = $1,651.

The cost of all printers is $645.

$1,651 − $645 = $1,006

The school spent $1,006 less on printers than on computers.

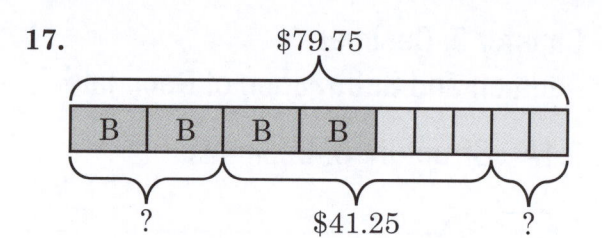

17.

2 books and 2 pens cost:
$79.75 − $41.25 = $38.50

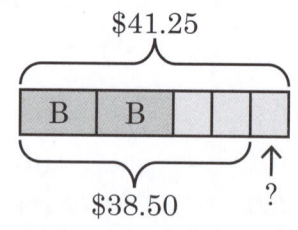

1 pen costs: $41.25 − $38.50 = $2.75

1 pen costs $2.75.

Multiplication of Decimals

3.2A Decimal Number System

Basics

1. **(a)** 0.05 **(b)** 0.5
   (c) 5   **(d)** 50
   (e) 5   **(f)** 0.5

2. **(a)** 400   **(b)** 4,200
   (c) 7.17 **(d)** 7.17
   (e) 900   **(f)** 0.009

Practice

3. **(a)** $48.50 × 10 = $485

   His older sister saved $485.

   (b) $48.50 − ($48.50 × $\frac{1}{10}$) = $43.65

   Eli has $43.65 left.

22 Chapter 3 DECIMALS

Challenge

4. Guitar: $78.50
   Piano: $78.50 × 100 = $7,850
   Violin: $7,850 × $\frac{1}{10}$ = $785
   $78.50 + $7,850 + $785 = $8,713.50
   She spent $8,713.50 on all three instruments.

3.2B Multiplier and Product

Basics

5. **(a)** The product will be less than 205.63 because the other factor is less than 1.
   (b) The product will be greater than 205.63 because the other factor is greater than 1.
   (c) The product will be less than 205.63 because the other factor is less than 1.
   (d) The product will be greater than 205.63 because the other factor is greater than 1.

Practice

6. (a) =   (b) >
   (c) <   (d) <
   (e) >   (f) <
   (g) =   (h) >
   (i) >

3.2C Multiplying Decimals

Basics

7. (a) 4.5   (b) 56.1
   (c) 0.35   (d) 0.0782
   (e) 13.12

Practice

8. **(a)** 42.5 in × 42.5 in = 1,806.25 in²
   The area is 1,806.25 square inches.
   (b) 2.7 m × 2.7 m × 2.7 m = 19.683 m³
   The volume is 19.683 cubic meters.
   (c) 3.2 × $18.27 = $58.464 ≈ $58.46
   Isaac will spend $58.46.

Challenge

9. 15.3 cm × 3.8 = 58.14 cm
   58.14 cm − 15.3 cm = 42.84 cm
   The shade-grown bean is 42.84 cm shorter than the sun-grown bean.

Chapter 3 DECIMALS

10.

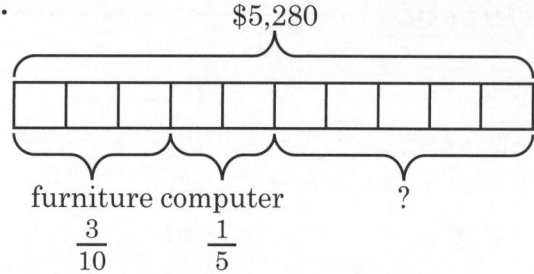

$5,280

furniture computer ?
$\frac{3}{10}$ $\frac{1}{5}$

10 units = $5,280

1 unit ⟶ $\frac{\$5,280}{10}$ = $528

5 units ⟶ $528 × 5 = $2,640

Consider the $2,640 as a new bar, of which we take a quarter.

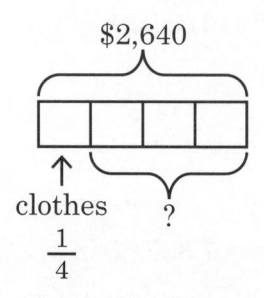

$2,640

clothes   ?
$\frac{1}{4}$

4 units = $2,640

1 unit ⟶ $660

$2,640 − $660 = $1,980

Alyssa had $1,980 left.

11.

$45.65

Week 1

Week 2   ?

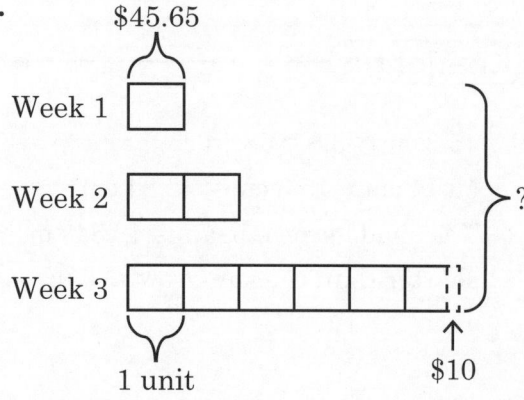

Week 3

1 unit $10

From the model,

1 unit = $45.65

9 units ⟶ 9 × $45.65 = $410.85

$410.85 − $10 = $400.85

Anna saved $400.85 altogether.

Division of Decimals

3.3A Divisor and Quotient

Basics

1. **(a)** The quotient will be less than the dividend because the divisor is greater than 1.
   (b) The quotient will be greater than the dividend because the divisor is less than 1.
   (c) The quotient will be less than the dividend because the divisor is greater than 1.
   (d) The quotient will be less than the dividend because the divisor is greater than 1.

2. **(a)** =   **(b)** <
   (c) <   **(d)** <
   (e) <   **(f)** >
   (g) >   **(h)** >
   (i) >

3.3B Dividing Decimals

Basics

3. **(a)** 3   **(b)** 7.2, 10
   (c) 6   **(d)** 3,090

4. **(a)** 8   **(b)** 0.8
   (c) 0.08   **(d)** 0.8
   (e) 800

Practice

5. **(a)** 0.5   **(b)** 1.5
   (c) 12.4   **(d)** 68.2
   (e) 32.5   **(f)** 2.6
   (g) 3.1

6. $\frac{20 \text{ tons}}{2.5} = 8$ tons
   At this rate, it will take 8 tons of gravel to improve each mile.

7. $\frac{330.7 \text{ m}^2}{15.6 \text{ m}} = 21.2$ m
   The width of the garden is 21.2 m.

8. $\frac{1{,}085 \text{ kg}}{51.7 \text{ kg}} = 21$
   He packaged 21 bags.

9. $\frac{\$17.94}{4.43 \text{ L}} = \$4.05$ per L
   $\frac{\$12.98}{2.95 \text{ L}} = \$4.40$ per L
   I should buy the jumbo size, which costs $4.05 per liter. The smaller size costs $4.40 per liter.

Challenge

10. 13.75 kg ÷ 2.75 = 5 kg. Each meter of pipe weighs 5 kg.
    2 m of pipe: 2 × 5 kg = 10 kg
    10 × $16.50 = $165.00
    2 meters of this pipe would cost $165.

3.4 Metric Measurements and Decimals

Basics

1. (a) 15,000   (b) 1,500
   (c) 150   (d) 15,000
   (e) 7,000   (f) 700
   (g) 62,000   (h) 6,200

2. (a) 0.005   (b) 0.0052
   (c) 0.013   (d) 0.0138
   (e) 0.0008   (f) 0.0088
   (g) 10.602   (h) 1.2

Practice

3. 8.3 m − 5.9 m = 2.4 m = 240 cm
   $\frac{240 \text{ cm}}{80 \text{ cm}} = 3$
   Hazel can make 3 cushions.

4. (a) 8.5 kg + 7.35 kg + 0.98 kg = 16.83 kg
   All 3 boxes weigh 16.83 kg.
   (b) 8.5 kg − 0.98 kg = 7.52 kg = 7,520 g
   The difference in weight between Box A and Box C is 7,520 g.

5. 4.6 L × 3 = 13.8 L
   13.8 L ÷ 24 = 0.575 L = 575 mL
   She should put 575 mL in each of the small containers.

6. 1.8 kg × 4 = 7.2 kg
   10 kg − 7.2 kg = 2.8 kg = 2,800 g
   The maximum weight for the fifth box is 2,800 g.

Challenge

7.

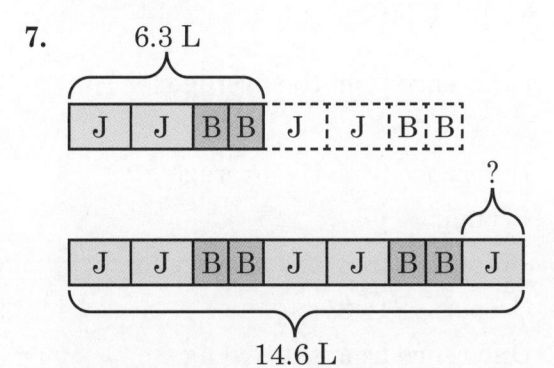

From the model,
2 jars and 2 bottles hold 6.3 L.
4 jars and 4 bottles = 2 × 6.3 L
= 12.6 L
14.6 L − 12.6 L = 2 L
1 jar holds 2 L, so 2 jars hold 4 L.
6.3 L − 4 L = 2.3 L
2 bottles can hold 2.3 L.
2.3 L ÷ 2 = 1.15 L
1 bottle can hold 1.15 L.

Chapter 4: Negative Numbers

4.1 Positive and Negative Numbers

Basics

1. **(a)** sea level, −3,000 m
   (b) 0 °F, +85 °F

2. 20 m above sea level

3. **(a)** −4   **(b)** 2

Practice

4. Difference from the average for bird A: 0.05 g
   Difference from the average for bird B: 0.02 g
   Difference from the average for bird C: −0.01 g
   Difference from the average for bird D: −0.005 g
   Difference from the average for bird E: 0.001 g

5. Difference from the average for Ada: −$6.87
   Difference from the average for Briana: $2.08
   Difference from the average for Cooper: −$18.85
   Difference from the average for Darryl: $15.37
   Difference from the average for Ethan: $8.27

6. Difference from the average for 49 Palms Oasis: −2.34 miles
   Difference from the average for Split Rock Loop: −2.84 miles
   Difference from the average for Lost Palms Oasis: 2.16 miles
   Difference from the average for Lost Horse Loop: 1.16 miles
   Difference from the average for Willow Hole: 1.86 miles

7. **(a)** 600 m − 443 m = 157 m
   The Canton Tower is 157 meters taller than the Willis Tower.
   (b) Using the Canton Tower as a reference point, the relative position of the man on the Willis Tower is −157 meters.

Challenge

8. **(a)** Using the Roosevelt Island Station as a reference point, the relative position of the commuter is −40 feet.
   (b) Using the 190th Street Station as a reference point, the relative position of the commuter at the 191st Street Station is −40 feet.
   (c) The distance between the Roosevelt Island or the Lexington Avenue at 63rd Street stations and the 191st Street station offers the greatest distance. Any of these three stations could be the reference point.

Comparing Positive and Negative Numbers

4.2A The Number Line

Basics

1. (a) [number line showing $-2\frac{1}{2}$ and 2.5 marked between -6 and 3]

   (b) $-5, -2\frac{1}{2}, 0, 2.5$

   (c) Answers will vary.
   Example answers: $-5, -6$

2. (a) [vertical number line with $1\frac{1}{2}$, $\frac{2}{1}$, $1\frac{1}{5}$, -2.4, -3.2 marked]

   (b) $-3.2, -2.4, 1\frac{1}{5}, 1\frac{1}{2}$

3. (a) > (b) <
   (c) >

4. (a) 2,675 m   (b) -10 °F
   (c) $-\$1{,}503$

Practice

5. Difference from the average for Ani: 6
   Difference from the average for Shanice: -2
   Difference from the average for Nolan: 5
   Difference from the average for Tyler: -8
   Difference from the average for Mia: -10

6. Difference from the average for Jan: $2\frac{1}{12}$
   Difference from the average for Feb: $\frac{5}{6}$
   Difference from the average for Mar: $\frac{1}{2}$
   Difference from the average for Apr: 0
   Difference from the average for May: $\frac{1}{3}$
   Difference from the average for Jun: -1
   Difference from the average for Jul: $-2\frac{1}{2}$
   Difference from the average for Aug: $-2\frac{7}{24}$
   Difference from the average for Sep: $-1\frac{2}{3}$

Challenge

7. $6, 4\frac{1}{2}, 0.4, -0.4, -4.25, -4\frac{1}{3}$

8. Answers will vary.
   Example solutions: $1.5, -1\frac{1}{2}$

4.2B Absolute Value

Basics

9. The the height of Willis Tower is 443 m.

10. (a) 4   (b) 4
    (c) 6   (d) 6
    (e) 8

11. (a) <   (b) >
    (c) <   (d) >
    (e) =   (f) >

Practice

12. (a) >   (b) <
    (c) >   (d) <
    (e) =   (f) <

13. $|-0.8|, |1|, |7.2|, |-72.3|, |73.2|$

14. |21|, |12|, |−12|, |−11|, |8|, |−8|

    Note that |12| and |−12| as well as |8| and |−8| can be shown in the opposite order.

Challenge

15. **(a)** 50 points   **(b)** −20 points

    (c) 8 correct, 2 unanswered

    (d) 8 correct, 2 incorrect

16. **(a)** −4   **(b)** 8

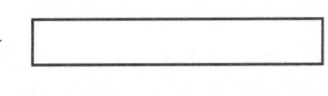

Chapter 5: Ratios
Ratios and Equivalent Ratios

5.1A Finding Ratio

Basics

1. (a) $7:2$ (b) $2:1$
    (c) $2:7$ (d) $1:2:7$
    (e) $7:10$
    (f) Answers will vary.
    Example answer: What is the ratio of clouds to hearts and stars? $1:9$

2. (a) $3:8$ (b) $8:3$
    (c) $3:11$

3. (a) $3:2$ (b) $2:3$
    (c) $5:2$

4. (a) $7:8$ (b) $8:15$

Practice

5. 3 yards = 9 feet
    $9:2$

6. 3 years 1 month = 37 months
    5 years = 60 months
    $37:60$

7. (a) $13:30$ (b) $3:43$

Challenge

8. Before
    John's money
    Amy's money

    After
    John's money — $8.75
    Amy's money
    $29.25

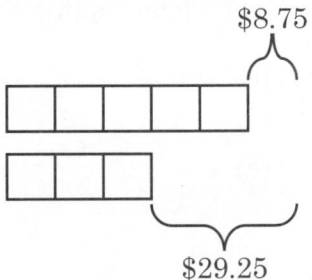

    $29.25 − $8.75 = $20.50

    From the model,
    2 units ⟶ $20.50
    1 unit ⟶ $20.50 ÷ 2 = $10.25
    3 units ⟶ 3 × $10.25 = $30.75
    $30.75 + $29.25 = $60.00
    Amy had $60.00 at first.

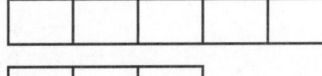

9. 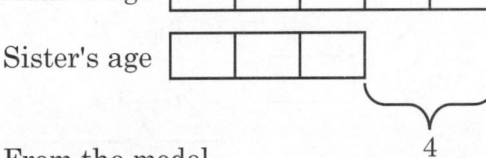 Adam's age
    Sister's age
    4

    From the model,
    2 units ⟶ 4
    1 unit ⟶ 2
    5 units ⟶ 10
    3 units ⟶ 6

    At this time, Adam is 10 years old and his sister is 6 years old.
    Adam's age in 5 years: 10 + 5 = 15
    Sister's age in 5 years: 6 + 5 = 11
    The ratio will be $15:11$.

5.1B Equivalent Ratios

Basics

10. (a) 4 : 3   (b) 3 : 4

11. (a) matches to (g), (b) matches to (e),
    (c) matches to (h), (d) matches to (f)

12. (a) 1 : 3   (b) 2 : 3
    (c) 1 : 4   (d) 4 : 7 : 1
    (e) 4 : 3   (f) 8 : 9
    (g) 7 : 5   (h) 2 : 6 : 5

13. (a) 9   (b) 8
    (c) 8   (d) 10, 5

Practice

14. (a) 2 : 5   (b) 5 : 7

15.

Lincoln walked 20 miles.

Amelia walked 20 mi − 12 mi = 8 mi.

Jett walked 20 mi + 8 mi = 28 mi.

The ratio of the distance Lincoln walked to the distance Amelia walked to the distance Jett walked was 5 : 2 : 7.

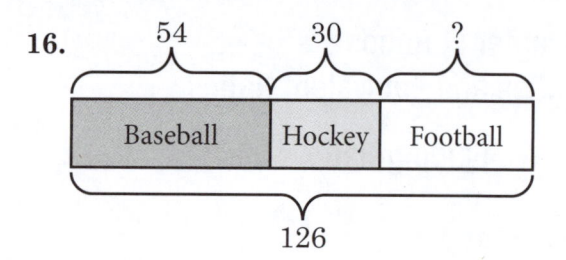

16.

$126 - 54 - 30 = 126 - 84 = 42$

Franco has 42 football cards.

$42 : 54 : 30 = 7 : 9 : 5$

17. (a) 78 : 27 or 26 : 9   (b) 27 : 78 or 9 : 26
    (c) 26 : 9   (d) No

18. If the shortest side is 12 cm, then the remaining sides are 16 cm and 20 cm, respectively.

The perimeter of the triangle is 48 cm.

Challenge

19. 12 m

From the model,

2 units → 12 m

1 unit → $\frac{12 \text{ m}}{2}$ = 6 m

9 units → 9 × 6 m = 54 m

The total length of the two pieces of ribbon is 54 m.

20. 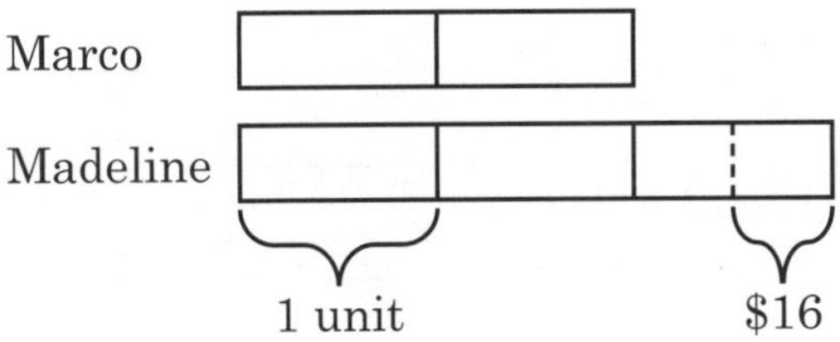 Marco
    Madeline
    1 unit $16

$16 must be half a unit, so that adding it to Marco and removing it from Madeline makes their amounts the same.

From the model,
1 unit → 2 × $16 = $32
2 units → 2 × $32 = $64
Marco has $64.

21. Before
    girls
    ?
    boys

    After
    girls
    6
    boys

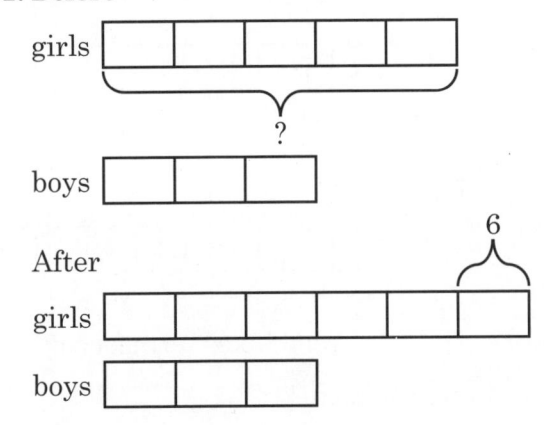

From the model,
1 unit → 6
5 units → 5 × 6 = 30
There were 30 girls in the club at first.

5.2 Ratios and Fractions

Basics

1. (a) $\frac{1}{4}$ (b) $\frac{2}{3}$
   (c) $\frac{14}{5}$ (d) $\frac{5}{2}$

2. The length of Ribbon A is $\frac{5}{4}$ the length of Ribbon B.

3. 3 : 5

Practice

4. (a) Alana got 7 units out of 16 units.
   $7 \div 16 = \frac{7}{16}$
   Alana got $\frac{7}{16}$ of the stamps.

   (b) Jerry got 9 units out of 16 units.
   $9 \div 16 = \frac{9}{16}$
   Jerry got $\frac{9}{16}$ of the stamps.

   (c)
   Alana
           ?
   Jerry
           27 stamps

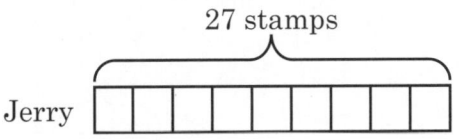

   From the model,
   9 units → 27
   1 unit → $\frac{27}{9}$ = 3
   7 units → 7 × 3 = 21
   Alana got 21 stamps.

5. (a) $\frac{3}{8}$ of the eating utensils are forks.
   (b) $\frac{5}{8}$ of the eating utensils are spoons.
   (c) Method 1
   Forks   15
   Spoons   ?

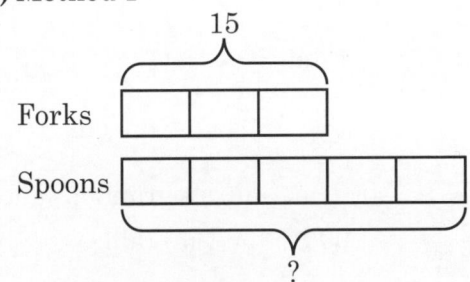

   From the model,
   3 units → 15
   1 unit → $\frac{15}{3}$ = 5
   5 units → 5 × 5 = 25

   Method 2
   Since there are $\frac{5}{3}$ as many spoons as forks, we multiply $\frac{5}{3}$ by 15.
   $\frac{5}{\cancel{3}_1} \times \cancel{15}^5 = 25$
   There are 25 spoons in the drawer.

6. Paula
   Kawai
   $312

   From the model,
   13 units → $312
   1 unit → $\frac{$312}{13}$ = $24
   6 units → 6 × $24 = $144
   7 units → 7 × $24 = $168
   Paula got $144. Kawai got $168.

7. 12 : 2 : 1 : 16

8. (a) 5 : 9
   (b)
   Granddaughter
   Mr. Nakamura
   180 cm

   Method 1
      From the model,
      9 units → 180 cm
      1 unit → $\frac{180 \text{ cm}}{9}$ = 20 cm
      5 units → 5 × 20 cm = 100 cm
   Method 2
      Find $\frac{5}{9}$ of 180.
      $\frac{5}{\cancel{9}} \times \cancel{180}^{20}$ cm = 100 cm
   Mr. Nakamura's granddaughter
   is 100 cm (or 1 m) tall.

9. Santiago's cards
   Amanda's cards
   42
   h   b
   From the model,
   7 units → 42 cards
   1 unit → $\frac{42}{7}$ = 6 cards
   Amanda has 2 units of hockey cards
   and 3 units of baseball cards. Amanda
   has 6 more baseball than hockey cards.

Challenge

10. (a) Eli's age
    Jamal's age
    Avery's age

    From the model, Eli's age to
    Avery's age is $\frac{3}{6} = \frac{1}{2}$.

    (b) From the model,
    4 units → 28 years
    1 unit → $\frac{28 \text{ years}}{4}$ = 7 years
    3 units → 3 × 7 years = 21 years
    Avery is 21 years older than Eli.

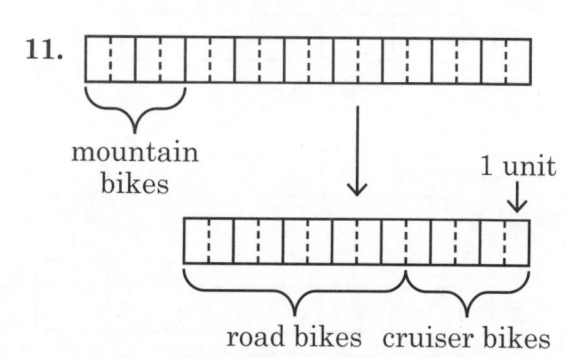

11.
    mountain bikes   1 unit

    road bikes   cruiser bikes
    From the model,
    cruiser bikes → 5 units
    mountain bikes → 4 units
    5 units − 4 units = 1 unit
    1 unit → 15 bikes
    9 units → 9 × 15 bikes = 135 bikes
    There are 135 road bikes in the
    bicycle shop.

12. Jenna's beads
    Koni's beads
    3 : 5 = 6 : 10
    The fraction is $\frac{5}{11}$.

32 Chapter 5 RATIOS

Chapter 6: Rate

6.1 Average and Rate

Basics

1. $9 + 8 + 6 + 9 = 32$
   $\frac{32}{4} = 8$
   Cooper's average was 8 points.
   He passed the course.

2. $42 + 78 + 81 + 23 + 12 = 236$
   $236 \div 5 = 47.2$

3. (a) $82 \text{ cm} + 64 \text{ cm} + 98 \text{ cm} + 44 \text{ cm}$
       $= 288 \text{ cm}$
       $288 \text{ cm} \div 4 = 72 \text{ cm}$
   (b) $3.25 \text{ L} + 7 \text{ L} + 2.5 \text{ L} + 8.45 \text{ L} + 1.75 \text{ L}$
       $= 22.95 \text{ L}$
       $22.95 \text{ L} \div 5 = 4.59 \text{ L}$

4. (a) $5.2 \text{ kg} + 3.5 \text{ kg} + 7 \text{ kg} + 9.74 \text{ kg}$
       $= 25.44 \text{ kg}$
       The total weight is 25.44 kg.
   (b) $25.44 \text{ kg} \div 4 = 6.36 \text{ kg}$
       The average weight is 6.36 kg.

5. $7.2 \text{ mi} + 5.5 \text{ mi} + 4 \text{ mi} + 8.7 \text{ mi} + 5 \text{ mi}$
   $= 30.4 \text{ mi}$
   $\frac{30.4 \text{ mi}}{5} = 6.08 \text{ mi}$
   Rounded to 1 decimal place, Darryl's average was 6.1 miles per day.

6. 1 day ⟶ 53 mi
   5 days ⟶ 53 mi × 5 = 265 mi
   They drove 265 miles on their vacation.

Practice

7. 24 ounces ⟶ 1 dog
   1 ounce ⟶ $\frac{1}{24}$ dog
   432 ounces ⟶ $\frac{\cancel{432}^{18}}{\cancel{24}_{1}} = 18$ dogs
   There is enough food for 18 dogs for 1 day.
   There are 18 large dogs in the kennel.

8. | 18.8 | 12.6 | 8.25 | 14 | ? |

   | 12.4 | 12.4 | 12.4 | 12.4 | 12.4 |

   Total = 5 × 12.4

   Total of 5 numbers: $5 \times 12.4 = 62$
   $18.8 + 12.6 + 8.25 + 14 = 53.65$
   $62 - 53.65 = 8.35$
   The fifth number is 8.35.

9. $75.50 ⟶ 1 week
   $528.50 ⟶ $\frac{\$528.50}{\$75.50} = 7$ weeks
   It will take Alexa 7 weeks to save $528.50.

10. $7.3 \text{ cm} \times 5 = 36.5 \text{ cm}$
    $5.3 \text{ cm} + 8.2 \text{ cm} + 9.1 \text{ cm} + 6.4 \text{ cm}$
    $= 29 \text{ cm}$
    $36.5 \text{ cm} - 29 \text{ cm} = 7.5 \text{ cm}$
    The missing seedling's length is 7.5 cm.

11. $287.40 \times 6 = \$1{,}724.40$
    $\$302.15 + \$243.65 + \$263.55 +$
    $\$305.45 + \$332.90 = \$1{,}447.70$
    $\$1{,}724.40 - \$1{,}447.70 = \$276.70$
    They spent $276.70 on gas in May.

12.

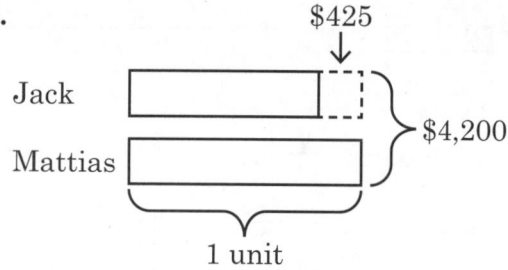

Jack / Mattias bar model: $425, $4,200, 1 unit

Average savings → $2,100
Total savings → $2,100 × 2 = $4,200
From the model,
2 units → $4,200 + $425 = $4,625
1 unit → $\frac{\$4,625}{2}$ = $2,312.50

Mattias has $2,312.50 in savings.

Challenge

13.

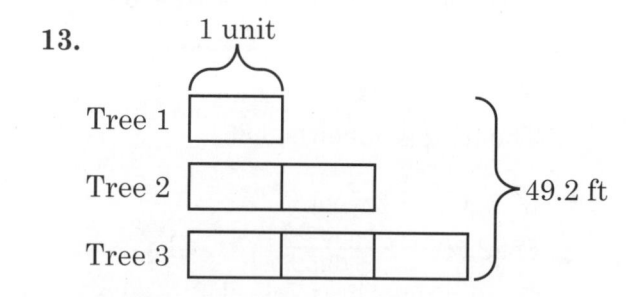

Tree 1 (1 unit), Tree 2 (2 units), Tree 3 (3 units); total 49.2 ft

Average height → 16.4 ft
Total height → 16.4 ft × 3 = 49.2 ft
1 unit → $\frac{49.2}{6}$ ft = 8.2 ft
2 units → 8.2 ft × 2 = 16.4 ft

The height of the second tree is 16.4 ft.

14. A + B = 87
B + C = 111
C − A = 24

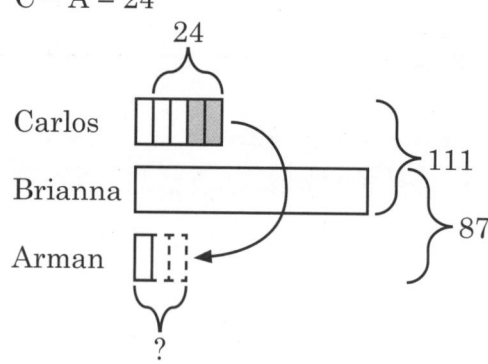

Carlos / Brianna / Arman bar model with 24, 111, 87

From the model,
4 units → 24
1 unit → 6
3 units → 18

The average number of stickers they have is 18.

6.2 Unit Rate

Basics

1. 45 minutes → 6.75 mi
1 minute → $\frac{6.75}{45}$ mi = 0.15 mi
She can run 0.15 miles per minute.

2. 60 minutes → 450 cans
1 minute → $\frac{450}{60}$ = 7.5 cans
It can pack 7.5 cans in one minute.

3. $1,650 ÷ 30 = $55.
It costs $55 per day during June.

4. 12 months → $15,420
1 month → $\frac{\$15,420}{12}$ = $1,285
3 months → $1,285 × 3
They pay $3,855 for three months' rent.

5. 252 puffs ⟶ 45 min

   1 puff ⟶ $\frac{45}{252}$ min

   448 puffs ⟶ $\frac{45}{252}$ min × 448

   $= \frac{5}{7}$ min × 112

   $= \frac{560}{7}$ min = 80 min

   It will take 80 minutes to fill 448 cream puffs.

Practice

6. 5 m ⟶ $12.35

   1 m ⟶ $12.35 ÷ 5 = $2.47

   2 m ⟶ 2 × $2.47 = $4.94

   2 m of wire cost $4.94.

7. 6,000 words ÷ 40 words per minute
   = 150 minutes

   6,000 words ÷ 75 words per minute
   = 80 minutes

   150 minutes − 80 minutes = 70 minutes

   It takes the faster typist 70 fewer minutes.

8. 1 hour ⟶ $48.50

   $2\frac{1}{2}$ hours ⟶ $48.50 × $\frac{5}{2}$ = $121.25

   $121.25 + $25 = $146.25

   He charged $146.25.

9. $3.85 × 17 = $65.45

   $65.45 + $8 = $73.45

   It will cost $73.45.

10. $18.80 × 40 = $752

    $18.80 × 1.5 = $28.20

    $28.20 × 12 = $338.40

    $752 + $338.40 = $1,090.40

    Andrei made $1,090.40.

11. (a) 20 min ⟶ 81 km

    1 min ⟶ $\frac{81}{20}$ km

    45 min ⟶ $\frac{81}{20}$ km × 45 = 182.25 km

    It can travel 182.25 km.

    (b) 81 km ⟶ 20 min

    1 km ⟶ $\frac{20}{81}$ min

    486 km ⟶ $\frac{20}{81}$ min × 486 = 120 min

    It would take 120 minutes or 2 hours.

Challenge

12. $12.50 × 2 = $25

    $8.75 × 4 = $35

    $7.50 × 4 = $30

    $25 + $35 + $30 = $90

    10 liters ⟶ $90

    1 liter ⟶ $\frac{\$90}{10}$ = $9

    3 liters ⟶ 3 × $9 = $27

    The cost is $27.

13. 900 pages ⟶ 15 min

    1 page ⟶ $\frac{15}{900}$ = $\frac{1}{60}$ min

    4,800 pages ⟶ $\frac{4,800}{60}$ = 80 min

    = 1 hr 20 min

    It will take 1 hr 20 min.

6.3 Speed

Basics

1. 5 hours ⟶ 56 mi
   1 hour ⟶ $\frac{56}{5}$ mi = 11.2 mi
   The average speed of the biker is 11.2 miles per hour.

2. 4.5 hours ⟶ 243 mi
   1 hour ⟶ $\frac{243}{4.5}$ mi = 54 mi
   Their average speed was 54 miles per hour.

3. 2.6 mi ⟶ 1 hr
   1 mi ⟶ $\frac{1}{2.6}$ hr
   6.5 mi ⟶ $\frac{1}{2.6}$ hr × 6.5 = 2.5 hr
   It will take Mrs. Jung 2.5 hours.

Practice

4. 49 mph × 3.75 hours = 183.75 mi
   It traveled 183.75 miles.

5. 45 minutes = $\frac{3}{4}$ hours
   $\overset{3}{\underset{2}{\cancel{\tfrac{3}{4}}}} \times \overset{7}{\cancel{14}}$ mi = $10\frac{1}{2}$ mi
   The distance between Ashimah's house and school is $10\frac{1}{2}$ miles.

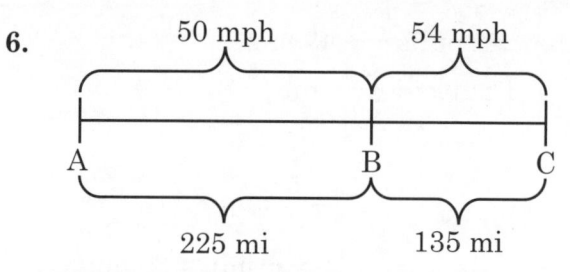

6.
   ```
         50 mph    54 mph
       ⌢──────⌢──────⌢
       A      B      C
       ⌣──────────⌣──
       225 mi   135 mi
   ```
   Time spent from City A to City B:
   $\frac{\text{Distance}}{\text{Rate}} = \frac{225 \text{ mi}}{50 \text{ mph}} = 4.5$ h

   Time spent from City B to City C:
   $\frac{\text{Distance}}{\text{Rate}} = \frac{135 \text{ mi}}{54 \text{ mph}} = 2.5$ h

   4.5 h + 2.5 h = 7 h
   It took the bus 7 hours to travel from City A to City C.

7. 217.35 km ÷ 2.25 h = 96.6 kph
   96.6 kph × 3 h = 289.8 km
   She will travel 289.8 km in 3 hours.

8. (a) 2.25 h × 70 mph = 157.5 mi
       It is 157.5 miles.
   (b) 157.5 mi ÷ 50 mph = 3.15 h
       It would take the truck 3.15 hours.

Challenge

9. 5 p.m. to 8:30 p.m. = 3.5 h
   1 h ⟶ 15 mi apart
   3.5 h ⟶ 3.5 × 15 mi = 52.5 mi
   Sumin was 52.5 miles from Town B at 8:30 p.m.

10. Ella traveled 90 mi in 90 minutes, which was $\frac{3}{5}$ of the journey.
    Distance between Town P and Town R:
    $5 \times \frac{90}{3} = 150$ mi
    Pablo traveled 50 mph for 150 miles.
    Time = $\frac{\text{Distance}}{\text{Rate}} = \frac{150}{50} = 3$ h
    He arrived at Town P at 1:00 p.m.

Chapter 7: Percent

7.1 Meaning of Percent

Basics

1. (a) $\frac{53}{100}$, 0.53, 53% of the shaded units
   (b) $\frac{150}{100} = \frac{15}{10} = 1\frac{1}{2}$, 1.5, 150% of the shaded units

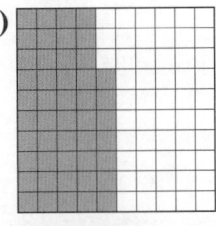

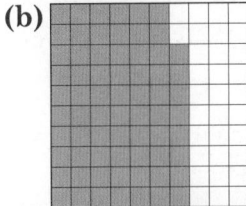

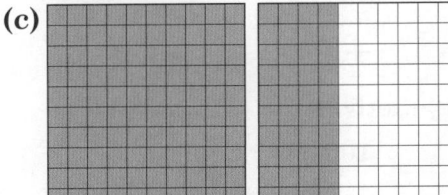

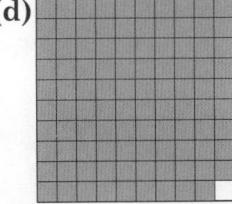

2. (a)
   (b)
   (c)
   (d)

Practice

3.

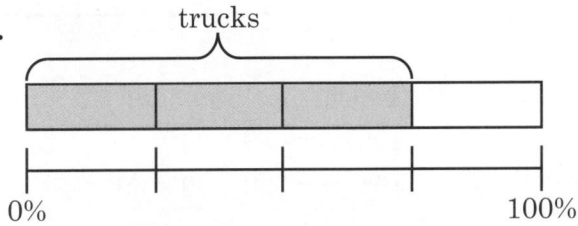

$\frac{3}{4} \times 100\% = 75\%$

75% of the vehicles in the parking lot are trucks.

4. $\frac{11}{20} \times 100\% = 55\%$

   55% of the dogs are beagles.

5. $\frac{8}{8} \longrightarrow 100\%$

   $\frac{1}{8} \longrightarrow 100\% \div 8 = 12.5\%$

   $\frac{8}{8} - \frac{3}{8} = \frac{5}{8}$

   $\frac{5}{8} \longrightarrow 5 \times 12.5\% = 62.5\%$

   She kept 62.5% of her work bonus.

6. Raj ran $1\frac{1}{2}$ times, or 150%, of the distance he intended to run.

7. (a) 66.6%   (b) 44.4%

8. (a) Already completed
   (b) 75%
   (c) 0.3; 30%   (d) $\frac{29}{50}$; 58%
   (e) $\frac{7}{200}$; 0.035   (f) $2\frac{1}{4}$; 2.25
   (g) $5\frac{1}{20}$; 5.05   (h) 4.28; 428%
   (i) $\frac{33}{75}$; 44%

9. $100\% - 60\% = 40\%$

   40% of the people at the concert are not children.

7.2 Percentage of a Quantity

Basics

1. **(a)** Method 1
   $100\% \longrightarrow \$90$
   $10\% \longrightarrow \$90 \div 10 = \$9$
   $60\% \longrightarrow 6 \times \$9 = \$54$
   Method 2
   60% of $90
   $= \frac{60}{100} \times \$90$
   $= \$54$
   60% of $90 is $54.

   (b) Method 1
   $100\% \longrightarrow \$90$
   $25\% \longrightarrow \$90 \div 4 = \$22.50$
   $75\% \longrightarrow 3 \times \$22.50 = \$67.50$
   Method 2
   75% of $90
   $= \frac{75}{100} \times \$90$
   $= \$67.50$
   75% of $90 is $67.50.

   (c) Method 1
   $100\% \longrightarrow \$90$
   $5\% \longrightarrow \$90 \div 20 = \$4.50$
   Method 2
   5% of $90
   $= \frac{5}{100} \times \$90$
   $= \$4.50$
   5% of $90 is $4.50.

   (d) Method 1
   $100\% \longrightarrow 180$ in
   $25\% \longrightarrow 180$ in $\div 4 = 45$ in
   Method 2
   25% of 180 in
   $= \frac{25}{100} \times 180$ in $= 45$ in

   25% of 180 inches is 45 inches.

   (e) $33\frac{1}{3}\% = \frac{\frac{100}{3}}{100}$
   $= \frac{100}{3} \times \frac{1}{100}$
   $= \frac{1}{3}$

   $33\frac{1}{3}\%$ of 105 km $= \frac{1}{3}$ of 105 km
   $= \frac{1}{3} \times 105$ km
   $= 35$ km

   $33\frac{1}{3}\%$ of 105 km is 35 km.

   (f) Method 1
   $100\% \longrightarrow 184$ cm
   $25\% \longrightarrow 184$ cm $\div 4 = 46$ cm
   $75\% \longrightarrow 3 \times 46$ cm $= 138$ cm
   Method 2
   75% of 184 cm
   $= \frac{75}{100} \times 184$ cm
   $= 138$ cm
   75% of 184 cm is 138 cm.

   (g) Method 1
   $100\% \longrightarrow \$200$
   $1\% \longrightarrow \$200 \div 100 = \$2$
   $8\frac{1}{2}\% \longrightarrow 8\frac{1}{2} \times \$2 = \$17$
   Method 2
   $8\frac{1}{2}\% = \frac{8.5}{100} \times \$200 = \$17$

   $8\frac{1}{2}\%$ of $200 is $17.

2. **(a)** $\frac{13}{25} \times 100\% = \frac{52}{100} \times 100\% = 52\%$

   (b) $\frac{105}{300} \times 100\% = \frac{7}{20} \times 100\% = 35\%$

3.

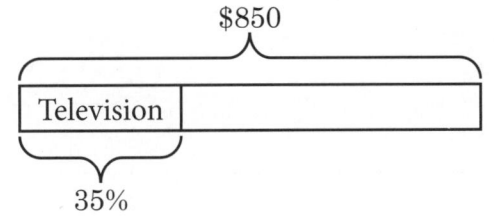

$\frac{35}{100} \times \$850 = 35 \times \$8.5 = \$297.50$

It cost $297.50.

4. 100% − 80% = 20%

20% × 65 = 13

13 pets in the store are not mammals.

Practice

5. $\frac{50 \text{ cm}}{10 \text{ m}} = \frac{50 \text{ cm}}{1,000 \text{ cm}} = \frac{5}{100} = 5\%$

50 cm is 5% of 10 m.

6. 75% of 97 = $\frac{75}{100} \times 97 = \$72.75$

80% of 88 = $\frac{80}{100} \times 88 = \$70.40$

The red blouse costs $2.35 less.

7. 43% ⟶ 387 trees

1% ⟶ $\frac{387}{43}$ = 9 trees

100% ⟶ 9 × 100 = 900 trees

There are 900 trees.

8. 100% − 38% = 62%

38% ⟶ 4.56 mi

1% ⟶ $\frac{4.56}{38}$ mi = 0.12 mi

62% ⟶ 62 × 0.12 mi = 7.44 mi

She hiked 7.44 miles after lunch.

9. 100% − 68% = 32%

32% ⟶ 192 games

1% ⟶ $\frac{192}{32}$ = 6 games

100% ⟶ 100 × 6 = 600 games

The store had 600 games at first.

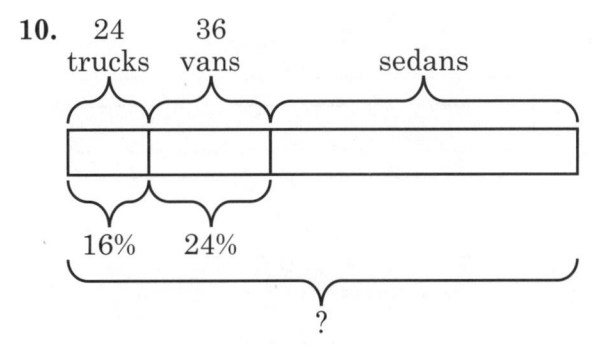

10. Trucks: 16% ⟶ 24

Trucks and vans: 40% ⟶ 60

1% ⟶ $\frac{60}{40}$ = 1.5

100% ⟶ 100 × 1.5 = 150

There are 150 vehicles.

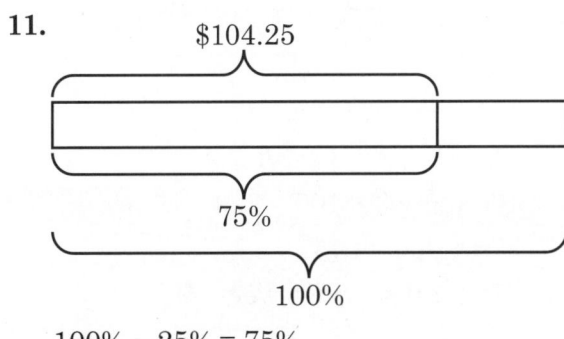

11. 100% − 25% = 75%

75% ⟶ $104.25

100% ⟶ $\frac{\$104.25}{75} \times 100 = \$34.75 \times 4$
    = $139

The original cost was $139.00.

12. 100% ⟶ $345.80

35% ⟶ $\frac{\$345.80}{100} \times 35 = \$121.03$

They saved $121.03.

13. 20% of 1,960 = 392 students

392 + 1,960 = 2,352 students

2,352 students are now attending the school.

14.

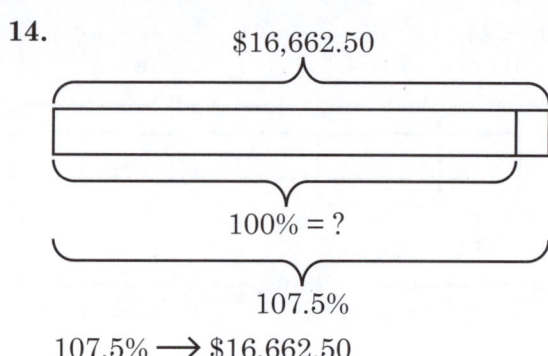

$16,662.50 — 107.5%
100% = ?

107.5% → $16,662.50

100% → $\frac{\$16,662.50}{107.5} \times 100$

= $155 × 100 = $15,500.00

The price was $15,500.00 before tax.

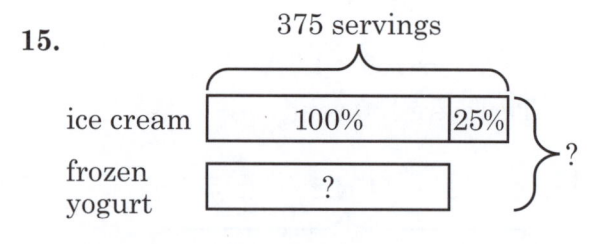

15.

375 servings

ice cream: 100% | 25%
frozen yogurt: ?

125% → 375 servings

1% → $\frac{375}{125}$ = 3 servings

100% → 3 × 100 = 300 servings

375 + 300 = 675

675 total servings were sold.

Challenge

16. First number:

25% → 125

1% → $\frac{125}{25}$ = 5

100% → 5 × 100 = 500

Second number:

35% → 10.5

1% → $\frac{10.5}{35}$ = 0.3

100% → 0.3 × 100 = 30

500 + 30 = 530

The sum of both numbers is 530.

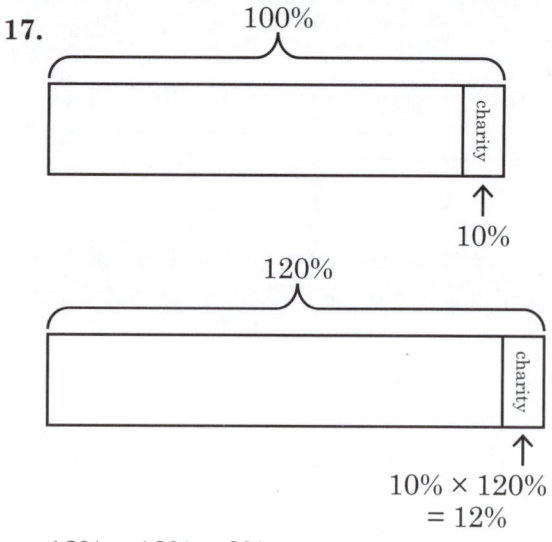

17.

100% — charity at 10%
120% — charity at 10% × 120% = 12%

12% − 10% = 2%

2% → $768

120% → $\frac{\$768}{2} \times 120 = \$768 \times 60$

= $46,080

Ms. Seville's income last year was $46,080.

18. Before

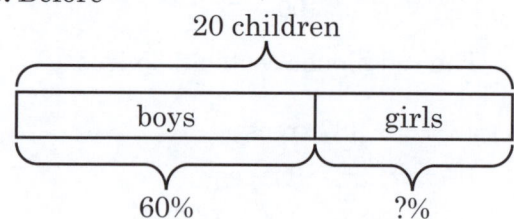

20 children

boys | girls
60% | ?%

Boys

100% → 20 children

60% → $\frac{60}{100} \times 20$ = 12 children

Girls

20 − 12 = 8 children

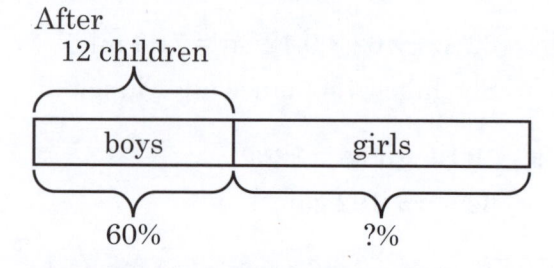

After

12 children

boys | girls
60% | ?%

40% → 12 children

10% → 12 ÷ 4 = 3 children

60% → 6 × 3 = 18 children

18 − 8 = 10 children

10 more girls got on the bus.

6B Solutions

Chapter 8: Algebraic Expressions
Writing and Evaluating Algebraic Expressions

8.1A Use of Letters

Basics

1. (a) $s - 5$ (b) $15\frac{1}{2} - k$
   (c) $h + 2.25$ (d) $100 + m$

2. (a) $\frac{5c}{6}$ (b) $8.2e$
   (c) $1{,}000r$ (d) mc^2
   (e) $\frac{t}{12}$

3. $\frac{s}{10}$ and $\frac{1}{10}s$ are equivalent because
   $\frac{s}{10} = \frac{1 \times s}{10} = \frac{1}{10}s$.

4. (a) $\frac{5k}{3}, \frac{5}{3}k$
   (b) $\frac{4c}{2}, \frac{4}{2}c$

5. (a) $3h - 5$
   (b) $\frac{a}{5} + 10$ or $\frac{1}{5}a + 10$
   (c) $\frac{3}{10}k + \frac{2}{5}$ or $\frac{3k}{10} + \frac{2}{5}$
   (d) $\frac{1}{2} - \frac{3}{5}t$ or $\frac{1}{2} - \frac{3t}{5}$
   (e) $7a + c$
   (f) $7n + 10p$

Practice

6. The difference is $(8 - k)$ minutes.

7. (a) $2p$ (b) $\frac{2}{3}p$
   (c) $p - 2$

8. Aki has $(23 - p)$ apps.

9. They will cost $(3b + 2p)$ dollars altogether.

10. Each friend got $\frac{c-2}{3}$ cookies.

11. They weigh $(3s + 2c)$ grams altogether.

Challenge

12. The area is w^3 cm.

13. 4 units $\longrightarrow 32b$
    1 unit $\longrightarrow 32b \div 4 = 8b$
    3 units $\longrightarrow 3 \times 8b = 24b$
    Santiago has $24b$ hockey cards.
    Santiago has $24b + n$ basketball cards.

8.1B Evaluating Algebraic Expressions

Basics

14. (a) When $y = 3$,
    $y + 10 = 3 + 10$
    $ = 13$
    (b) When $y = 3$,
    $5y = 5 \times 3$
    $ = 15$
    (c) When $y = 3$,
    $y^4 = 3 \times 3 \times 3 \times 3$
    $ = 81$
    (d) When $y = 3$,
    $\frac{4y}{6} = \frac{4 \times 3}{6} = \frac{12}{6}$
    $\phantom{\frac{4y}{6}} = 2$

15. (a) When $n = 10$,
$\frac{3}{5}n = \frac{3 \times 10}{5}$
$= \frac{30}{5}$
$= 6$

(b) When $n = 10$,
$1.2n = 1.2 \times 10$
$= 12$

(c) When $n = 10$,
$\frac{5}{3} \times 10 = \frac{5 \times 10}{3} = \frac{50}{3} = 16\frac{2}{3}$

(d) When $n = 10$,
$0.07n = 0.07 \times 10$
$= 0.7$

16. (a) When $h = 6$,
$7 + 3h = 7 + 3 \times 6$
$= 25$

(b) When $h = 6$,
$9h - 22 = 9 \times 6 - 22$
$= 32$

(c) When $h = 6$,
$(18 - 2h) \div 4 = (18 - 2 \times 6) \div 4$
$= \frac{6}{4} = 1\frac{1}{2}$

(d) When $h = 6$,
$18 - \frac{2h}{4} = 18 - \frac{2 \times 6}{4}$
$= 18 - 3 = 15$

(e) When $h = 6$,
$\frac{1}{3}h + 15 = \frac{6}{3} + 15$
$= 17$

Practice

17. (a) When $m = \frac{1}{2}$,
$6m - 2 = 6 \times \frac{1}{2} - 2$
$= 1$

(b) When $m = \frac{1}{2}$,
$5 + 4m = 5 + 4 \times \frac{1}{2}$
$= 7$

(c) When $m = \frac{1}{2}$,
$\frac{9}{14} - 4m \div 7 = \frac{9}{14} - 4 \times \frac{1}{2} \div 7$
$= \frac{9}{14} - \frac{2}{7}$
$= \frac{9}{14} - \frac{4}{14}$
$= \frac{5}{14}$

(d) When $m = \frac{1}{2}$,
$\frac{9}{14} - \frac{4m}{7} = \frac{9}{14} - 4 \times \frac{\frac{1}{2}}{7}$
$= \frac{9}{14} - \frac{2}{7}$
$= \frac{9}{14} - \frac{4}{14}$
$= \frac{5}{14}$

(e) When $m = \frac{1}{2}$,
$\frac{1}{2}m - \frac{1}{4} = \frac{1}{2} \times \frac{1}{2} - \frac{1}{4}$
$= 0$

18. (a) When $s = \frac{3}{4}$,
$\frac{s}{6} = \frac{3}{4} \times \frac{1}{6} = \frac{1}{8}$

(b) When $s = \frac{3}{4}$,
$\frac{2s}{8} = \frac{2}{8} \times s = \frac{2}{8} \times \frac{3}{4} = \frac{3}{16}$

(c) When $s = \frac{3}{4}$,
$\frac{s}{12} + 9 = \frac{1}{12} \times \frac{3}{4} + 9$
$= \frac{1}{16} + 9 = 9\frac{1}{16}$

(d) When $s = \frac{3}{4}$,
$\frac{8s}{5} - \frac{5}{6} = \frac{8}{5} \times \frac{3}{4} - \frac{5}{6}$
$= \frac{6}{5} - \frac{5}{6}$
$= \frac{36}{30} - \frac{25}{30}$
$= \frac{11}{30}$

19. **(a)** When $b = 0.45$,
$\frac{3b}{5} = \frac{3 \times 0.45}{5}$
$= \frac{1.35}{5} = 0.27$

(b) When $b = 0.45$,
$b^2 = 0.45 \times 0.45$
$= 0.2025$

(Challenge)

20. **(a)** When $a = \frac{2}{3}$, $b = \frac{1}{2}$, $c = \frac{3}{4}$,
$\frac{a}{b} \times 8c = (\frac{2}{3} \div \frac{1}{2}) \times 8 \times \frac{3}{4}$
$= \frac{2}{3} \times 2 \times 8 \times \frac{3}{4}$
$= 8$

(b) When $a = \frac{2}{3}$, $b = \frac{1}{2}$, $c = \frac{3}{4}$,
$3a + \frac{b}{8} - c^2$
$= 3 \times \frac{2}{3} + (\frac{1}{2} \div 8) - (\frac{3}{4} \times \frac{3}{4})$
$= 2 + \frac{1}{2} \times \frac{1}{8} - \frac{9}{16}$
$= 2 + \frac{1}{16} - \frac{9}{16}$
$= \frac{3}{2} = 1\frac{1}{2}$

8.1C Word Problems

(Basics)

21. **(a)**

Number of Arrangements	Total Number of Flowers
1	12
2	24
3	36
4	48
5	60

(b) $\frac{f}{12}$ arrangements could be made.

(c) $\frac{180}{12} = 15$

15 can be made.

(Practice)

22. **(a)** The shade-grown bean plant is $(b - 3)$ cm.

(b) If $b = 32$, then
$b - 3 = 32 - 3 = 29$
The shade-grown plant is 29 cm tall.

23. **(a)** $d + 75$

(b) $4 + 75 = 79$
The antique desk is 79 years old.

24. **(a)** $g \times 8.6$

(b) Its area is $8.6g$ m².

(c) 8.6 m $\times 7.8$ m $= 67.08$ m²
Its area is 67.08 m².

25. **(a)** There are $2.5c$ dogs.

(b) If $c = 4$, then $2.5c = 2.5 \times 4 = 10$.
There are 10 dogs.

26. **(a)** $w = \frac{7}{3}l$ or $2\frac{1}{3}l$

(b) If the width is 28 m, then $28 = 2\frac{1}{3}l$.
$28 \div \frac{7}{3} = l$
$28 \times \frac{3}{7} = 12$
The length of the field is 12 m.

Challenge

27. (a) The number of footballs is 7.5s.

The number of baseballs is 2.5s.

(b) If the number of footballs is 7.5s, then the number of soccer balls is $\frac{105}{7.5}$.

$\frac{105}{7.5} = 14$

There are 14 soccer balls.

28. The cost of one marker is $m = \frac{d-4p}{3}$.

29. Method 1

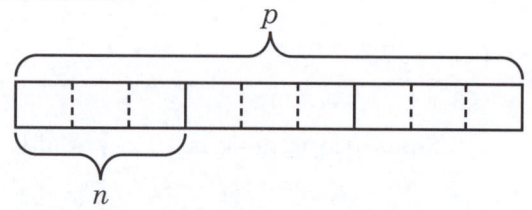

From the model,

9 units → p

1 unit → $\frac{p}{9}$

Method 2

$3n = p$

$n = \frac{p}{3}$

$\frac{1}{3}n = \frac{p}{9}$

8.2 Simplifying Algebraic Expressions

Basics

1. (a) $2(m + m) = 2m + 2m$
   $= 4m$

(b) $2m \times m = 2 \times m \times m$
   $= 2m^2$

(c) $8e - 2e = 6e$

2. (a) $(8s - 2s) + 3 = 6s + 3$

(b) $(y + y) + 28 = 2y + 28$

(c) $(8g - 5g) - 2 = 3g - 2$

(d) $(5h - 3h) + (4 - 1) = 2h + 3$

(e) $(3k - k) + (8 - 3) = 2k + 5$

(f) $10 - 2 - 5t - 2t = 8 - (5t + 2t) = 8 - 7t$

(g) $(4c - c) - c^2 + (32 + 18) = 3c - c^2 + 50$

3. (a) $\frac{9}{3}p = 3p$

(b) $\frac{3}{5} \times \frac{2}{3} \times k = \frac{2}{5}k$

(c) $4g \times 5 = 4 \times 5 \times g = 20g$

Practice

4. (a) The expressions are equivalent because of the Commutative Property of Addition.

(b) The expressions are not equivalent.

(c) Simplify both expressions to see if they are equal.

$5r + 4 + r - 2 = (5r + r) + (4 - 2)$
$= 6r + 2$

$1 + 7r + 1 - r = (7r - r) + (1 + 1)$
$= 6r + 2$

They are equivalent because both expressions simplify to $6r + 2$.

(d) Simplify both expressions to see if they are equal.

$7 + k + 2 + 3k = (7 + 2) + (k + 3k)$
$= 9 + 4k$

$3k + 5 + 4 + 2k = (3k + 2k) + (5 + 4)$
$= 5k + 9$

They are not equivalent because $9 + 4k \ne 5k + 9$.

5. (a) $(6 \times 3b) - (6 \times 4) = 18b - 24$

(b) $(7 \times 2e) + (7 \times 3) = 14e + 21$

(c) $\left(2 \times \frac{1}{3}\right) + (2 \times j) = \frac{2}{3} + 2j$

(d) $(15 \times 3) - (15 \times p) = 45 - 15p$

6. (a) $14y + 7 = 7 \times 2y + 7$
    $= 7 \times (2y + 1)$
    $= 7(2y + 1)$
   (b) $9x + 12 = 3 \times 3x + 3 \times 4$
    $= 3 \times (3x + 4)$
    $= 3(3x + 4)$
   (c) $15 - 10l = 5 \times 3 - 5 \times 2l$
    $= 5 \times (3 - 2l)$
    $= 5(3 - 2l)$
   (d) $16m - 4 = 4 \times 4m - 4$
    $= 4 \times (4m - 1)$
    $= 4(4m - 1)$
   (e) $32w + 8 - 4w - 4$
    $= 4 \times 8w + 4 \times 2 - 4 \times w - 4$
    $= 4 \times (8w + 2 - w - 1)$
    $= 4(8w - w + 2 - 1)$
    $= 4(7w + 1)$

Challenge

7. (a) $8 \times a + 8 \times b - 2 \times 3a - 2 \times 2b$
    $= 8a + 8b - 6a - 4b$
    $= 2a + 4b$
   (b) $(3c + 3d) - c^3 + 2c$
    $= 3c + 2c + 3d - c^3$
    $= 5c + 3d - c^3$

8. Yara's mother's age now is $4n \times 8$.
   $4n \times 8 = 32n$ years old
   5 years ago, Yara's mother's age was
   $(32n - 5)$ years old.

9. (a) $\$8m + 2(\$m + \$100) = \$10m + \$200$
    $= \$(10m + 200)$
   (b) $\$10m + \$200 = \$250 + \$200 = \$450$
    $= \$1{,}000 - \$450 = \$550$
    She has $550 left.

10. $\frac{1}{2}$ lb = 8 oz
    8 oz $\longrightarrow \$n$
    1 oz $\longrightarrow \$\frac{n}{8}$
    2 oz $\longrightarrow 2 \times \$\frac{n}{8} = \$\frac{n}{4}$ or $\$\frac{1}{4}n$

Chapter 9: Equations and Inequalities
Equations

9.1A Algebraic Equations

Basics

1. Begin by substituting 4 for r in each equation. Then, solve each equation to determine if it is true.

   (a) $4 + \frac{3}{4} = \frac{16}{4} + \frac{3}{4} = \frac{19}{4}$

   $\frac{19}{4} = \frac{19}{4}$

   Both sides of the equation are equal. Thus, $r = 4$ is a solution to this equation.

   (b) $49 = 45 + r$

   $45 + 4 = 49$

   $49 = 49$

   Both sides of the equation are equal. Thus, $r = 4$ is a solution to this equation.

   (c) $88 = 92 - r$

   $88 = 92 - 4$

   $88 = 88$

   Both sides of the equation are equal. Thus, $r = 4$ is a solution to this equation.

   (d) $16 - r = \frac{12}{3}$

   $16 - 4 = \frac{12}{3}$

   $12 \neq 4$

   Both sides of the equation are not equal. Thus, $r = 4$ is not a solution to this equation.

2. Begin by substituting 9 for c in each equation. Then, solve each equation to determine if it is true.

   (a) $3c = 27$

   $3 \times 9 = 27$

   $27 = 27$

   Thus, $c = 9$ is a solution to this equation.

   (b) $4 = \frac{3}{4}c$

   $4 = \frac{3}{4} \times 9$

   $4 \neq 6\frac{3}{4}$

   Thus, $c = 9$ is not a solution to this equation.

   (c) $63 = 7c$

   $63 = 7 \times 9$

   $63 = 63$

   Thus, $c = 9$ is a solution to this equation.

   (d) $11.7 = 1.3c$

   $11.7 = 1.3 \times 9$

   $11.7 = 11.7$

   Thus, $c = 9$ is a solution to this equation.

3. Begin by substituting $\frac{2}{5}$ for m in each equation. Then, solve each equation to determine if it is true.

   (a) $m + m = \frac{4}{10}$

   $\frac{2}{5} + \frac{2}{5} = \frac{4}{5}$

   $\frac{4}{5} \neq \frac{4}{10}$

   Thus, $m = \frac{2}{5}$ is not a solution to this equation.

   (b) $15m = 6$

   $15 \times \frac{2}{5} = \frac{30}{5}$

   $\frac{30}{5} = 6$

   $6 = 6$

   Thus, $m = \frac{2}{5}$ is a solution to this equation.

(c) $40 = 20m$
$20 \times \dfrac{2}{5} = \dfrac{40}{5}$
$\dfrac{40}{5} = 8$
$40 \neq 8$
Thus, $m = \dfrac{2}{5}$ is not a solution to this equation.

(d) $\dfrac{4}{15} = \dfrac{2}{3}m$
$\dfrac{4}{15} = \dfrac{2}{3} \times \dfrac{2}{5}$
$\dfrac{4}{15} = \dfrac{4}{15}$
Thus, $m = \dfrac{2}{5}$ is a solution to this equation.

(e) $20m - 4 = 10m$
$20 \times \dfrac{2}{5} - 4 = 4$
$10 \times \dfrac{2}{5} = 4 = 4$
$4 = 4$
Thus, $m = \dfrac{2}{5}$ is a solution to this equation.

4. Begin by substituting 0.32 for d in each equation. Then, solve each equation to determine if it is true.

(a) $1 - 2d = 0.64$
$1 - 2 \times 0.32 = 0.36$
$0.36 \neq 0.64$
Thus, $d = 0.32$ is not a solution to this equation.

(b) $d^2 = 0.1024$
$0.32 \times 0.32 = 0.1024$
$0.1024 = 0.1024$
Thus, $d = 0.32$ is a solution to this equation.

(c) $1 - d^2 \times 2 = 0.4754 + d$
$1 - 0.32 \times 0.32 \times 2 = 1 - 2 \times 0.1024$
$= 0.7952$
$0.4754 + 0.32 = 0.7954$
$0.7952 \neq 0.7954$
Thus, 0.32 is not a solution to this equation.

9.1B Balancing Equations

Basics

5. (a) $q + 9 = 27$
$q + 9 - 9 = 27 - 9$
$q = 18$
Check
$18 + 9 = 27$

(b) $p + 3.2 = 9$
$p + 3.2 - 3.2 = 9 - 3.2$
$p = 9 - 3.2$
$p = 5.8$
Check
$5.8 + 3.2 = 9$

(c) $y + \dfrac{3}{15} = \dfrac{4}{5}$
$y + \dfrac{3}{15} - \dfrac{3}{15} = \dfrac{4}{5} - \dfrac{3}{15}$
$y = \dfrac{12}{15} - \dfrac{3}{15}$
$y = \dfrac{9}{15} = \dfrac{3}{5}$
Check
$\dfrac{9}{15} + \dfrac{3}{15} = \dfrac{12}{15}$
$= \dfrac{4}{5}$

(d) $t + 16.8 = 55$
$t + 16.8 - 16.8 = 55 - 16.8$
$t = 38.2$
Check
$38.2 + 16.8 = 55$

6. (a) $p - 7 = 21$
$p - 7 + 7 = 21 + 7$
$p = 28$
Check
$28 - 7 = 21$
(b) $z - 93 = 128$
$z - 93 + 93 = 128 + 93$
$z = 221$
Check
$221 - 93 = 128$
(c) $y - \frac{3}{4} = \frac{3}{8}$
$y - \frac{3}{4} + \frac{3}{4} = \frac{3}{8} + \frac{3}{4}$
$y = \frac{3}{8} + \frac{6}{8}$
$y = \frac{9}{8}$ or $1\frac{1}{8}$
Check
$\frac{9}{8} - \frac{3}{4} = \frac{9}{8} - \frac{6}{8}$
$= \frac{3}{8}$
(d) $s - 0.56 = 2.93$
$s - 0.56 + 0.56 = 2.93 + 0.56$
$s = 3.49$
Check
$3.49 - 0.56 = 2.93$
(e) $40 - v = 12$
$40 - v + v = 12 + v$
$40 = 12 + v$
$40 - 12 = 12 - 12 + v$
$28 = v$
$v = 28$
Check
$40 - 28 = 12$

7. (a) $100 = n + 37$
$100 - 37 = n + 37 - 37$
$63 = n$
$n = 63$
Check
$63 + 37 = 100$
(b) $3 = y + \frac{6}{5}$
$3 - \frac{6}{5} = y + \frac{6}{5} - \frac{6}{5}$
$1\frac{4}{5} = y$
$y = 1\frac{4}{5}$
Check
$1\frac{4}{5} + \frac{6}{5} = 3$
(c) $\frac{5}{8} = n - 3$
$\frac{5}{8} + 3 = n - 3 + 3$
$3\frac{5}{8} = n$
$n = 3\frac{5}{8}$
Check
$3\frac{5}{8} - 3 = \frac{5}{8}$
(d) $1.84 = p + 0.98$
$1.84 - 0.98 = p + 0.98 - 0.98$
$0.86 = p$
$p = 0.86$
Check
$0.86 + 0.98 = 1.84$

8. **(a)** $6n = 84$

$\dfrac{6n}{6} = \dfrac{84}{6}$

$n = 14$

Check

$6 \times 14 = 84$

(b) $3n = 8.4$

$\dfrac{3n}{3} = \dfrac{8.4}{3}$

$n = 2.8$

Check

$3 \times 2.8 = 8.4$

(c) $4.2k = 25.2$

$\dfrac{4.2k}{4.2} = \dfrac{25.2}{4.2}$

$k = 6$

Check

$4.2 \times 6 = 25.2$

(d) $1.5s = 45$

$\dfrac{1.5s}{1.5} = \dfrac{45}{1.5}$

$s = 30$

Check

$1.5 \times 30 = 45$

9. **(a)** $\dfrac{x}{5} \times 5 = 7 \times 5$

$x = 35$

Check

$\dfrac{35}{5} = 7$

(b) $\dfrac{e}{7} \times 7 = 12 \times 7$

$e = 84$

Check

$\dfrac{84}{7} = 12$

(c) $\dfrac{m}{3} \times 3 = 1.02 \times 3$

$m = 3.06$

Check

$\dfrac{3.06}{3} = 1.02$

(d) $\dfrac{x}{4} \times 4 = 11.2 \times 4$

$x = 44.8$

Check

$\dfrac{44.8}{4} = 11.2$

Practice

10. **(a)** $3 \times \dfrac{2}{3}m = 3 \times 12$

$2m = 36$

$\dfrac{2m}{2} = \dfrac{36}{2}$

$m = 18$

Check

$\dfrac{2}{3} \times 18 = \dfrac{36}{2} = 12$

(b) $\dfrac{7}{8}e \div 7 = 4.2 \div 7$

$\dfrac{1}{8}e = 0.6$

$\dfrac{1}{8}e \times 8 = 0.6 \times 8$

$e = 4.8$

Check

$\dfrac{7}{8} \times 4.8 = \dfrac{7 \times 4.8}{8}$

$= \dfrac{7 \times 0.6}{1} = 4.2$

(c) $\dfrac{4}{3} \times \dfrac{3}{4}r = \dfrac{4}{3} \times 75$

$r = \dfrac{4 \times 75}{3}$

$r = \dfrac{300}{3}$

$r = 100$

Check

$\dfrac{3}{4} \times 100 = \dfrac{300}{4} = 75$

(d) $5 \times \dfrac{2}{5}d = 5 \times \dfrac{1}{2}$

$2d = \dfrac{5}{2}$

$2d \times \dfrac{1}{2} = \dfrac{5}{2} \times \dfrac{1}{2}$

$d = 1\dfrac{1}{4}$

Check

$\dfrac{2}{5} \times 1\dfrac{1}{4} = \dfrac{2}{5} \times \dfrac{5}{4} = \dfrac{1}{2}$

(e) $\dfrac{7}{9}b + 3 - 3 = 66 - 3$

$\dfrac{7}{9}b = 63$

$\dfrac{7}{9}b \times \dfrac{9}{7} = 63 \times \dfrac{9}{7}$

$b = 9 \times 9$

$b = 81$

Check

$\dfrac{7}{9} \times 81 + 3 = 63 + 3 = 66$

11. Method 1

Let c represent the number of baseball cards Isaac had before his birthday.

Algebraically:

$c + 29 = 413$

$c + 29 - 29 = 413 - 29$

$c = 384$

Method 2

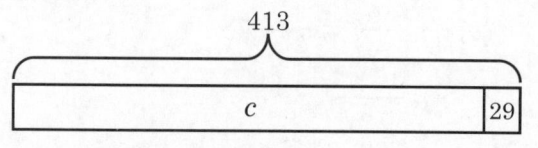

413	
c	29

From the model,

$c = 413 - 29$

$c = 384$

He had 384 baseball cards before his birthday.

12. Method 1

Algebraically, let b represent the number of math tests corrected in the morning.

$b + 17 = 43$

$b + 17 - 17 = 43 - 17$

$b = 26$

Method 2

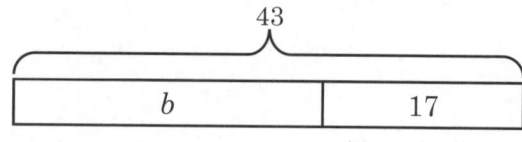

43	
b	17

From the model,

$b = 43 - 17$

$b = 26$

There were 26 math tests corrected in the morning.

13. Method 1

Algebraically, let f represent the amount of fabric in yards that Ellen had before she made the dress.

$f - 2\dfrac{2}{3} = 3\dfrac{5}{8}$

$f - 2\dfrac{2}{3} + 2\dfrac{2}{3} = 3\dfrac{5}{8} + 2\dfrac{2}{3}$

$f = 3\dfrac{15}{24} + 2\dfrac{14}{24}$

$f = 5\dfrac{31}{24} = 6\dfrac{7}{24}$

Method 2

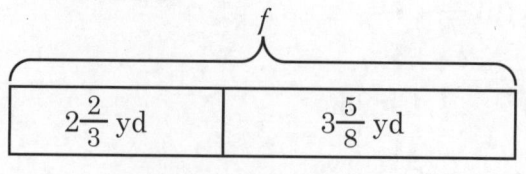

	f	
$2\dfrac{2}{3}$ yd		$3\dfrac{5}{8}$ yd

From the model,

$f = 2\dfrac{2}{3} + 3\dfrac{5}{8}$

$f = 2\dfrac{16}{24} + 3\dfrac{15}{24}$

$f = 5\dfrac{31}{24} = 6\dfrac{7}{24}$

Ellen had $6\dfrac{7}{24}$ yards of fabric before she made the dress.

14. Method 1

Algebraically, let m represent the amount of milk in cups that Melvin had at first.

$m - 4\frac{2}{3} = 2\frac{3}{4}$

$m - 4\frac{2}{3} + 4\frac{2}{3} = 2\frac{3}{4} + 4\frac{2}{3}$

$m = 2\frac{9}{12} + 4\frac{8}{12}$

$m = 6\frac{17}{12} = 7\frac{5}{12}$

Method 2

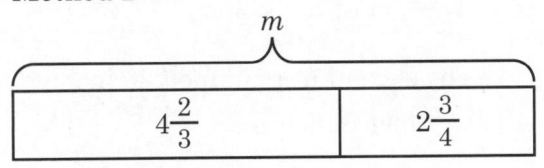

m	
$4\frac{2}{3}$	$2\frac{3}{4}$

From the model,

$m = \left(4\frac{2}{3} + 2\frac{3}{4}\right)$

$m = 7\frac{5}{12}$

He had $7\frac{5}{12}$ cups of milk.

15. Method 1

Algebraically, let y represent Rea's age in years.

$\frac{2}{3}y = 12$

$3 \times \frac{2}{3}y = 3 \times 12$

$2y = 36$

$\frac{2y}{2} = \frac{36}{2}$

$y = 18$

Method 2

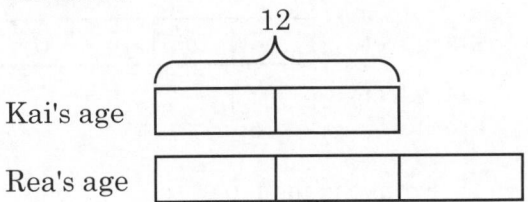

Kai's age / Rea's age (bar model with 12 over Rea's first two units)

$\frac{2}{3}y = 12$

$\frac{1}{3}y = 12 \div 2 = 6$

$\frac{3}{3}y = 3 \times 6 = 18$

Rea is 18 years old.

16. Method 1

Algebraically, let b represent amount of money Margaret's brother earned babysitting.

$$4b = 78$$
$$\frac{4b}{4} = \frac{78}{4}$$
$$b = 19.50$$

Method 2

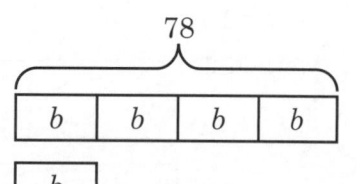

Margaret	b	b	b	b

Margaret's brother: b

Total: 78

From the model,
$$4b = 78$$
$$b = \frac{78}{4}$$
$$b = 19.50$$

Margaret's brother made $19.50 babysitting.

17. Method 1

Algebraically, let j be the distance the second frog jumped. Then $\frac{3}{4}j$ represents the distance the first frog jumped.

3 ft 6 in = 42 in

$$\frac{3}{4}j = 42$$
$$\frac{1}{4}j = 42 \div 3 = 14$$
$$\frac{4}{4}j = 14 \times 4 = 56$$

Method 2

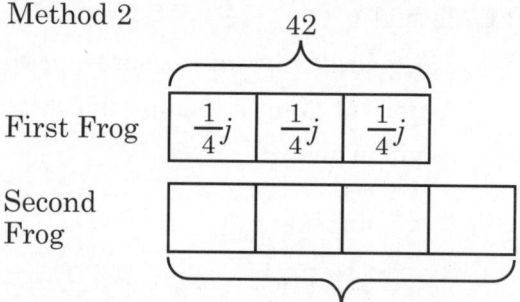

First Frog: $\frac{1}{4}j$ | $\frac{1}{4}j$ | $\frac{1}{4}j$ (total 42)

Second Frog: 4 parts (total j)

$$\frac{3}{4}j = 42$$
$$\frac{1}{4}j = 42 \div 3 = 14$$
$$\frac{4}{4}j = 14 \times 4 = 56$$

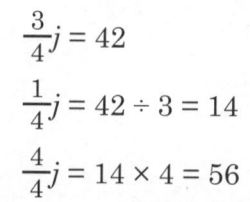

The second frog jumped 56 inches, or 4 feet 8 inches.

18. Method 1

Algebraically, let s represent the height of the small plant.

$$3s = 16.5$$
$$\frac{3s}{3} = \frac{16.5}{3}$$
$$s = 5.5$$

Method 2

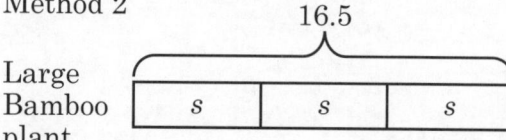

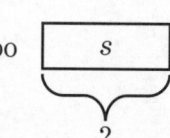

Large Bamboo plant: s | s | s (total 16.5)

Small Bamboo plant: s

$$3s = 16.5$$
$$s = \frac{16.5}{3}$$
$$s = 5.5$$

The small bamboo plant is 5.5 feet tall.

19. Method 1

Algebraically, let d be the distance, in miles, that Mr. Fujimoto drove.

$$d + \frac{1}{3}d = 348$$
$$1\frac{1}{3}d = 348$$
$$\frac{4}{3}d = 348$$
$$\left(\frac{3}{4} \times \frac{4}{3}\right)d = \frac{3}{4} \times 348$$
$$d = 3 \times 87 = 261$$

Method 2

Distance Mr. Fujimoto drove / Distance Mrs. Fujimoto drove = $\frac{1}{3}d$; total 348

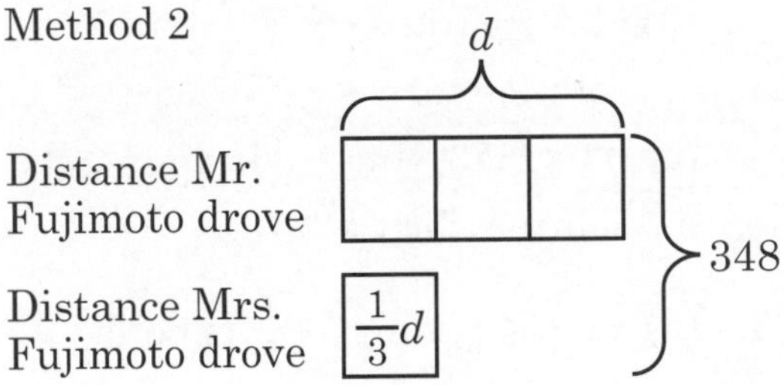

From the model,
$$d + \frac{1}{3}d = 348$$
$$1\frac{1}{3}d = 348$$
$$\frac{4}{3}d = 348$$
$$d = 261$$

Mr. Fujimoto drove about 261 miles.

20. (a) Paula's age = y
Arman's age = $y + 5$
Ms. Madani's age = $4(y + 5)$
$\qquad = 4y + 20$

(b) $4(y + 5) = 32$
$4y + 20 = 32$
$4y + 20 - 20 = 32 - 20$
$4y = 12$
$\frac{4y}{4} = \frac{12}{4}$
$y = 3$

Paula is 3 years old.

21. Method 1

Algebraically, let c be the number of caps they knitted in September.
$2c + 26 = 250$
$2c + 26 - 26 = 250 - 26$
$2c = 224$
$\frac{2c}{2} = \frac{224}{2}$
$c = 112$

Method 2

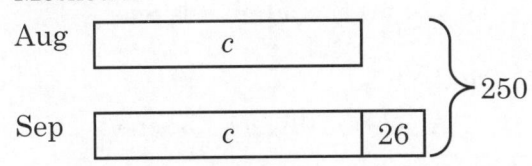

Aug: c
Sep: c, 26; total 250

From the model,
$2c = 250 - 26 = 224$
$c = \frac{224}{2}$
$c = 112$

The club knitted 112 caps in September.

22. $x + x + 30 = 180$
$2x + 30 - 30 = 180 - 30$
$2x = 150$
$2x \div 2 = 150 \div 2$
$x = 75$

Challenge

23. Method 1

Algebraically,
$\frac{1}{4}n = h$
$4 \times \frac{1}{4}n = 4 \times h$
$n = 4h$
$2n = 2 \times 4h = 8h$

Method 2

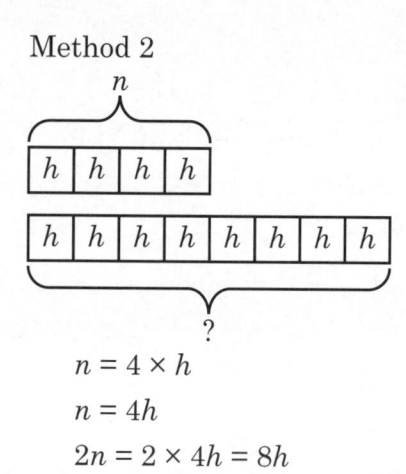

| h | h | h | h |

| h | h | h | h | h | h | h | h |

$n = 4 \times h$

$n = 4h$

$2n = 2 \times 4h = 8h$

Two times the number is $8h$.

24. Method 1

Algebraically, let x be the number of nickels.

$0.05x + 0.10(x + 8) = 2.00$

$5x + 10x + 80 = 200$

$15x + 80 - 80 = 200 - 80$

$15x = 120$

$\dfrac{15x}{15} = \dfrac{120}{15}$

$x = 8$

Method 2

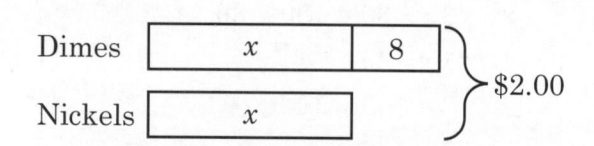

Dimes | x | 8 |
Nickels | x |

$\}$ $2.00

From the model,

$0.05x + 0.10(x + 8) = 2.00$

$5x + 10x + 80 = 200$

$15x + 80 - 80 = 200 - 80$

$15x = 120$

$15x = \dfrac{120}{15}$

$x = 8$

Ani has 8 nickels in her coin collection.

Inequalities
9.2A Algebraic Inequalities

Basics

1. (a) $4 > -3$, thus 4 is a solution for $x > -3$.
   (b) $4 < 4\frac{1}{8}$, thus 4 is a solution for $x < 4\frac{1}{8}$.
   (c) $4 = 4$, thus 4 is not a solution for $x < 4$.

2. (a) $-3 > -5.06$, thus -3 is a solution for $y > -5.06$.
   (b) $-3 = -3$, thus -3 is not a solution for $y < -3$.
   (c) $-3 < 3.73$, thus -3 is not a solution for $y > 3.73$.

3. (a) $-1\frac{1}{5} < 1$, thus $-1\frac{1}{5}$ is not a solution for $x > 1$.
   (b) $-1\frac{1}{5} > -2$, thus $-1\frac{1}{5}$ is not a solution for $x < -2$.
   (c) $-1\frac{1}{5} < -1$, thus $-1\frac{1}{5}$ is not a solution for $x > -1$.

9.2B Graphing Inequalities Using a Number Line

Basics

4. (a) [number line with open circle at 1, arrow extending right, marks from -3 to 3]

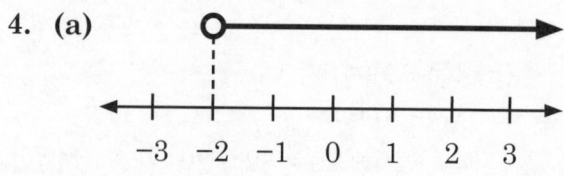

   (b) [number line with closed circle at -1, arrow extending right, marks from -3 to 4]

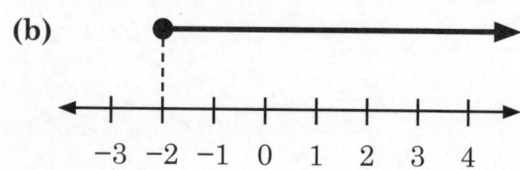

   (c) [number line with open circle at -4, arrow extending left, marks from -7 to -1]

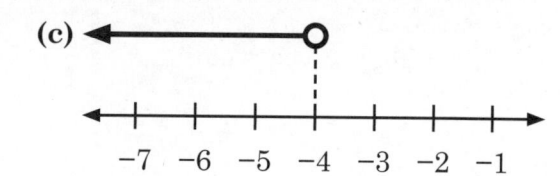

(d) [number line showing closed circle at 8, arrow extending left; marks at 6, 7, 8, 9, 10, 11, 12, 13]

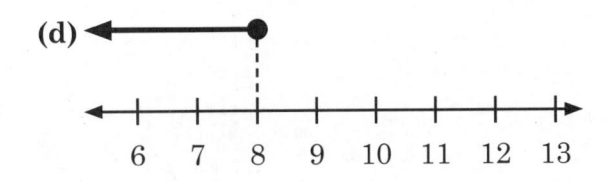

(e) [number line showing closed circle at $-2\frac{1}{2}$, arrow extending right; marks at $-3, -2\frac{1}{2}, -2, -1\frac{1}{2}, -1, -\frac{1}{2}, 0, 1$]

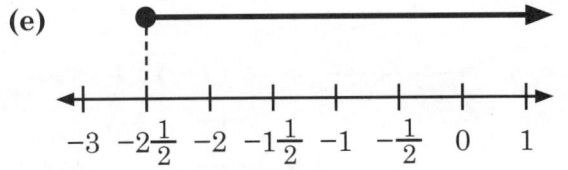

(f) [number line showing open circle at 3.5, arrow extending left; marks at 2, 2.5, 3, 3.5, 4, 4.5, 5]

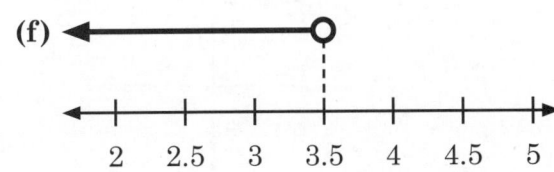

5. **(a)** $x \geq 800$, where x is the amount of money in dollars.

   (b) [number line with closed circle at 800, arrow extending right; marks at 0, 200, 400, 600, 800, 1,000, 1,200]

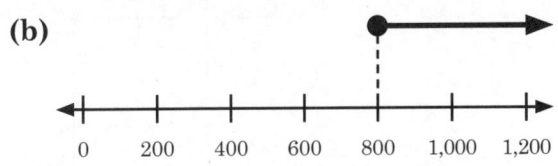

6. **(a)** $d \geq 4$, where d is Daniel's height in feet.

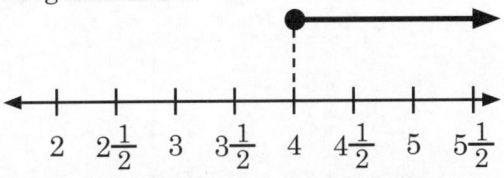

   [number line with closed circle at 4, arrow extending right; marks at $2, 2\frac{1}{2}, 3, 3\frac{1}{2}, 4, 4\frac{1}{2}, 5, 5\frac{1}{2}$]

   (b) $0 < s \leq 18$, where s is Samuel's amount of money in dollars.

   [number line with open circle at 0 and closed circle at 18; marks at 0, 2, 4, 6, 8, 10, 12, 14, 16, 18, 20]

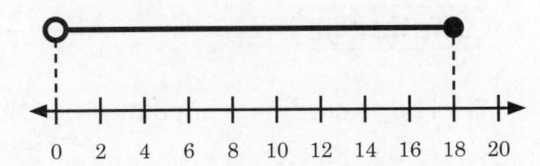

   (c) $h > 6.5$, where h is the weight of the grapes Harry has in lbs.

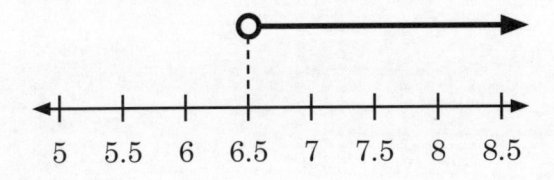

   [number line with open circle at 6.5, arrow extending right; marks at 5, 5.5, 6, 6.5, 7, 7.5, 8, 8.5]

(d) $45 \leq s \leq 65$, where s is the legal speed for the highway.

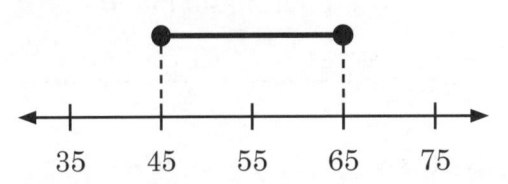

[number line with closed circles at 45 and 65; marks at 35, 45, 55, 65, 75]

Practice

7. **(a)** Answers will vary. All answers must be whole numbers greater than or equal to 2.

   (b) $h \geq 2$, where h is the number of hours spent volunteering by each student.

   (c) [number line with closed circle at 2, arrow extending right; marks at 0, 1, 2, 3, 4, 5]

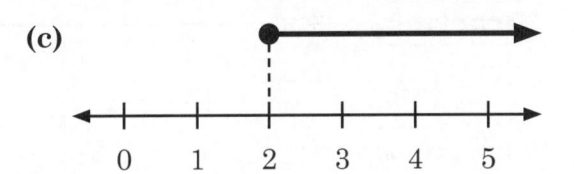

Challenge

8. $16h \geq 128$

   $$\frac{16h}{16} \geq \frac{128}{16}$$

   $h \geq 8$, where h is the number of hours Linda spends walking dogs. Linda needs to work at least 8 hours.

9. $3 \leq t \leq 4.5$

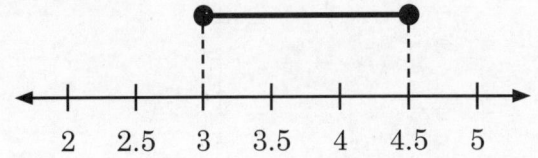

   [number line with closed circles at 3 and 4.5; marks at 2, 2.5, 3, 3.5, 4, 4.5, 5]

Chapter 10: Coordinates and Graphs

10.1 The Coordinate Plane

Basics

1. (a) 3   (b) −2
   (c) 7   (d) 0

2. (a) −1.75   (b) 0
   (c) 8   (d) $-\frac{1}{2}$

3. (a) 2nd Quadrant   (b) 1st Quadrant
   (c) 3rd Quadrant   (d) 4th Quadrant
   (e) 3rd Quadrant   (f) 4th Quadrant

Practice

4. V: (2, 2.5)   W: (0, 1)
   X: (−4, 1)   Y: (−2, −2)
   Z: (4, −1)

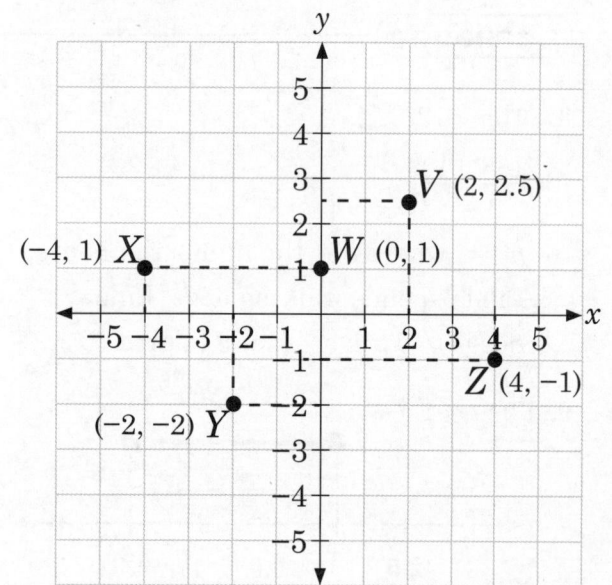

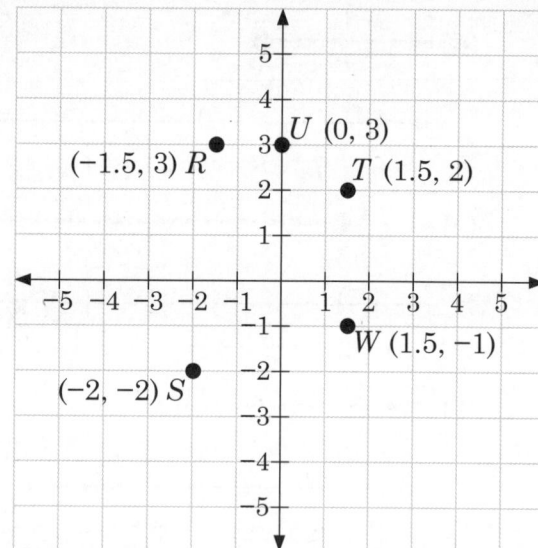

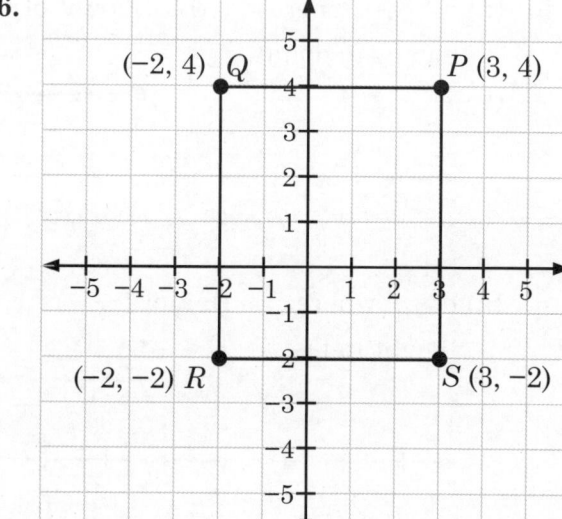

5. (see graph)

6. The polygon is a rectangle.

Challenge

7. The y-coordinate of point F is either −1 or −3.

10.2 Distance between Coordinate Pairs

Basics

1. (a) $|15| = 15$
   (b) $|-8| - |-2| = 6$
   (c) $|5| + |-1| = 6$
   (d) $|-6.2| + |5| = 11.2$

2. (a) $A(4, 8)$ and $B(4, -3)$ form a vertical line segment, not in the same quadrant
   (b) $C(-2, 3)$ and $D(-4, 3)$ form a horizontal line segment, in the same quadrant, 2nd quadrant
   (c) $E(-1, 2)$ and $F(-1, -8)$ form a vertical line segment, not in the same quadrant
   (d) $G(1, -5)$ and $H(-5, -5)$ form a horizontal line segment, not in the same quadrant

3. (a) $P(-1, 3)$ and $Q(-8, 3)$
       Yes, 7
   (b) $R(-1, -5)$ and $S(-6, -5)$
       Yes, 5
   (c) $T(3, -2)$ and $U(6, -2)$
       Yes, 3
   (d) $V(-1\frac{1}{2}, 6)$ and $W(1\frac{1}{2}, 6)$
       No, 3

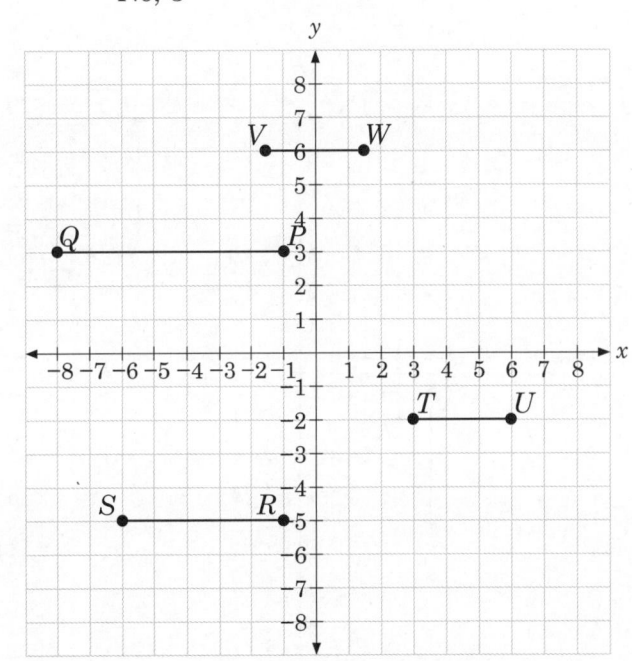

4. (a) H(2, −1) and I(2, −4)
   Yes, 3
   (b) J(−3, 4) and K(−3, −2)
   No, 6
   (c) L(5, 7) and M(5, 4)
   Yes, 3
   (d) N(−4, −1) and O(−4, −5)
   Yes, 4

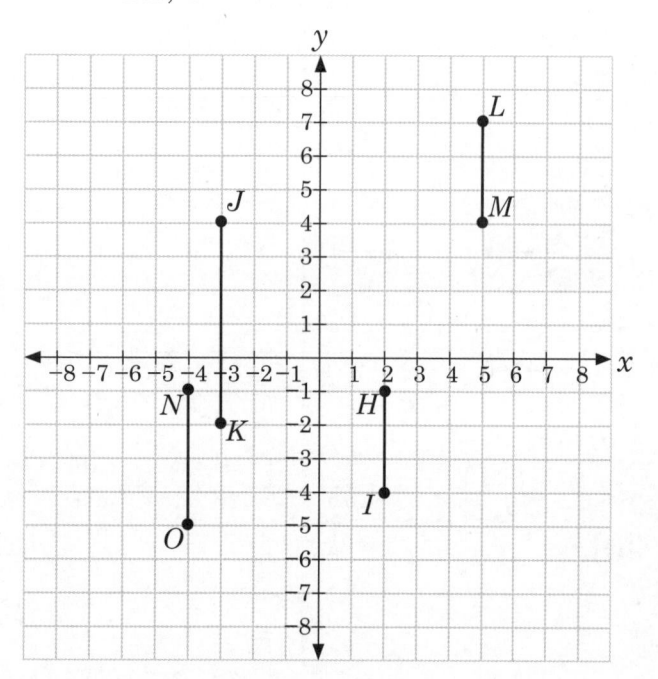

5. The distance is 11 meters.

6. The distance is 8 centimeters.

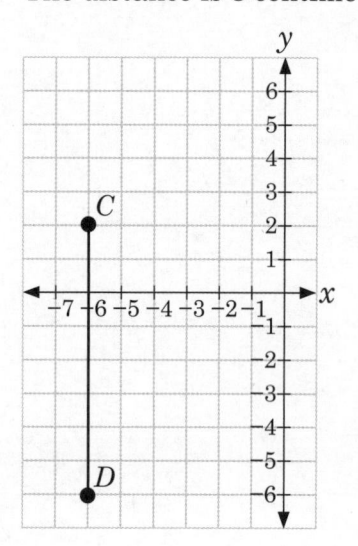

Practice

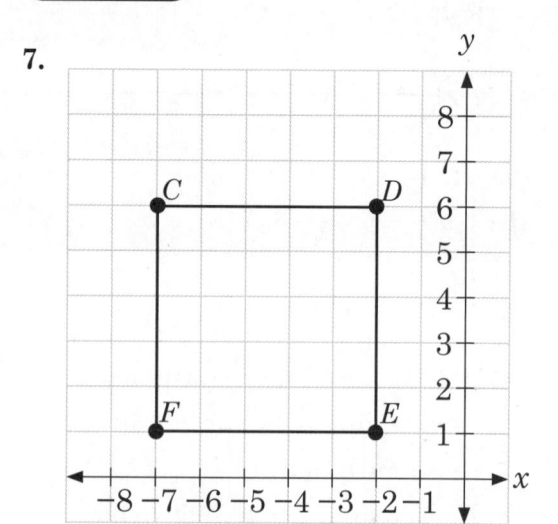

7. Area = 25 cm²
   Perimeter = 20 cm

8. (a)

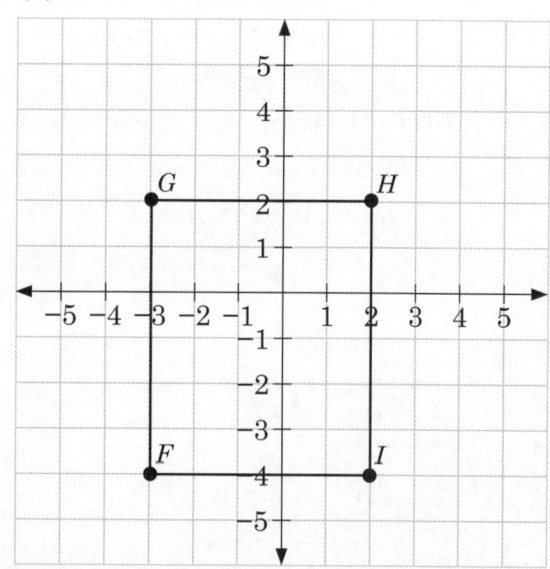

   (b) Length: |2| cm + |−4| cm = 2 cm + 4 cm = 6 cm
   Width: |2| cm + |−3| cm = 2 cm + 3 cm = 5 cm
   2(5 + 6) cm = 22 cm
   The perimeter of rectangle FGHI is 22 cm.

   (c) 6 cm × 5 cm = 30 cm²
   The area of rectangle FGHI is 30 cm².

9.

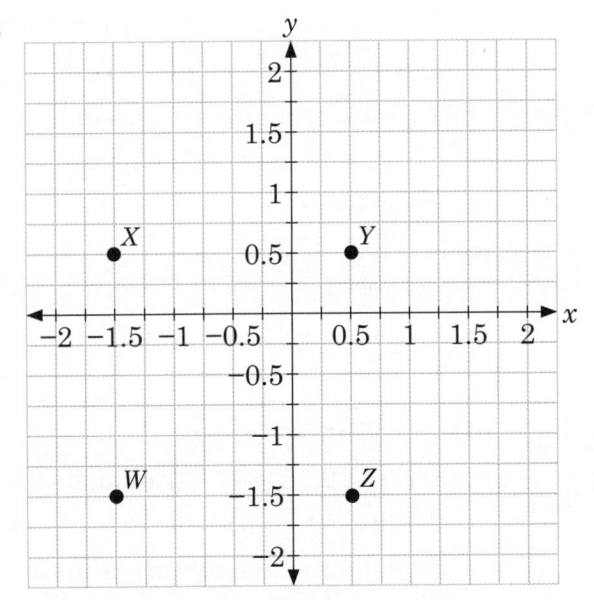

The coordinates of point Z are (0.5, −1.5).

10. **(a)** This forms a vertical line, because both *x*-coordinates are the same.
 9 m − 4 m = 5 m
 The length of the line is 5 m.
 (b) This forms a horizontal line because both *y*-coordinates are the same.
 |−45| m − |−15| m = 30 m
 The length of the line is 30 m.
 (c) This forms a vertical line, because both *x*-coordinates are the same.
 |2.5| m + |−5.5| m = 8 m
 The length of the line is 8 m.
 (d) This forms a horizontal line, because both *y*-coordinates are the same.
 |−5| m + |4.5| m = 9.5 m
 This length of the line is 9.5 m.

11. **(a)** Horizontal, no
 (b) Vertical, yes
 (c) Horizontal, no
 (d) Horizontal, no

12. **(a)** |−7.5| − |−5| = 2.5 cm
 (b) $|-3\frac{3}{4}| - |-1| = 2\frac{3}{4}$ cm
 (c) |110| − |99| = 11 cm
 (d) |−125| − |−75| = 50 cm
 (e) $|89| - |85\frac{1}{2}| = 3\frac{1}{2}$ cm
 (f) $|-45\frac{5}{8}| - |-15\frac{1}{8}| = 30\frac{1}{2}$ cm

13. **(a)** The distance between Points *Y* and *Z* is 10 cm.
 (b) Either:
 $(6\frac{1}{2}, 2\frac{1}{4})$ and $(6\frac{1}{2}, -7\frac{3}{4})$
 or
 $(-13\frac{1}{2}, 2\frac{1}{4})$ and $(-13\frac{1}{2}, -7\frac{3}{4})$
 (c) Area: 100 cm²
 Perimeter: 40 cm

14. Area: 40 × 55 = 2,200 units²
 The area is 2,200 units².

 Perimeter: 2(40 + 55) = 190 units
 The perimeter is 190 units.

Challenge

15. The coordinates of the other three vertices are (4, −4), (−2, −4), and (−2, 4).

16. The 4 coordinates could be:
 (4, −5), (−4, −5), (−4, 3), and (4, 3)
 or: (5, −4), (−3, −4), (−3, 4), and (5, 4)

17. Base is *BC*, length 8.
 Height is *AX*, length 8.
 Area = $\frac{1}{2}bh$
 = $\frac{1}{2} \times 8 \times 8$
 = 32 units²
 The area of triangle ABC is 32 units².

Changes in Quantities

10.3A Independent and Dependent Variables

Basics

1. (a) Independent variable: number of quarts Jennifer picked, b
   Dependent variable: number of pints Jennifer can make, p
   (b) Independent variable: number of baskets Joyce made, b
   Dependent variable: number of points Joyce scored, p
   (c) Independent variable: number of chores Jessica, c
   Dependent variable: number of stickers Jessica earns, s

2.

	Independent variable	Dependent variable
(a)	h	b
(b)	m	c
(c)	g	w
(d)	h	c

10.3B Representing Relationships between Variables

Basics

3. (a)

Grandfather's Age G	55	56	57	58	59	60
Amy's Age A	3	4	5	6	7	8

   (b) $G = A + 52$ or $A = G - 52$
   (c) $A = 75 - 52 = 23$
   Amy will be 23 years old when her grandfather is 75.

4. (a) The number of hours Carter works, h, is the independent variable. The total amount of money Carter earns, t, is the dependent variable.
   (b) $t = \$15.25h$
   (c)

Hours worked h	8	9	10	11	12
Total earned t (\$)	122	137.25	152.50	167.75	183

   (d) $\frac{3}{5} \times \$15.25 = \$9.15$
   $\$300 \div \$9.15 = 32.8$ (rounded to the nearest tenth)
   Carter will have to work almost 33 hours to save \$300.

5. (a) $t = 2.60 + 0.46m$
   (b) The independent variable is the number of miles. The dependent variable is the total cost of the cab ride.
   (c)

Miles m	Total cost t (\$)
10	7.20
11	7.66
12	8.12
13	8.58
14	9.04
15	9.50

   (d) $\$2.60 + 16 \times \$0.46 = \$9.96$
   It will cost \$9.96.

10.3C Observing Relations between Variables with Graphs

Basics

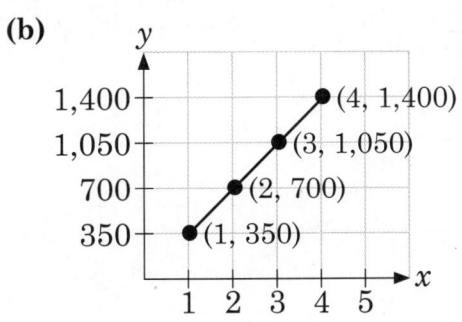

6. **(a)** $m = 350d$
   (b)

   Graph showing points (1, 350), (2, 700), (3, 1050), (4, 1400).

7. **(a)** The independent variable is the number of hours Albert drives, h. The dependent variable is the number of miles he drives, m.

   (b)

Hours h	3	4	5	6	7	8	9	10
Miles m	180	240	300	360	420	480	540	600

   (c) $m = 60h$

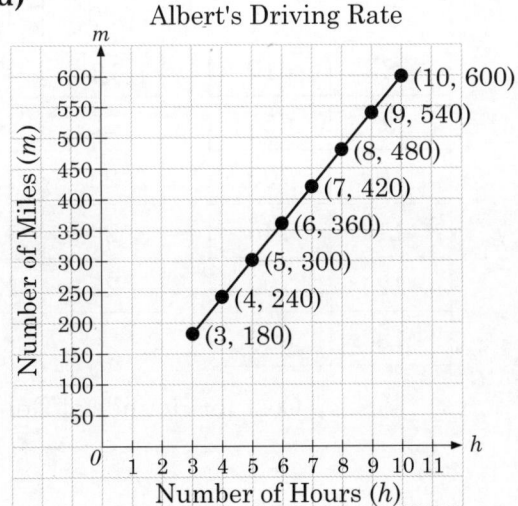

   (d) Albert's Driving Rate

   Graph with points (3, 180), (4, 240), (5, 300), (6, 360), (7, 420), (8, 480), (9, 540), (10, 600).

8. **(a)** $y = 39x$
   (b) Answers will vary depending on the values chosen. Example solution:

Bolts x	5	6	7	8	9
Yards y	195	234	273	312	351

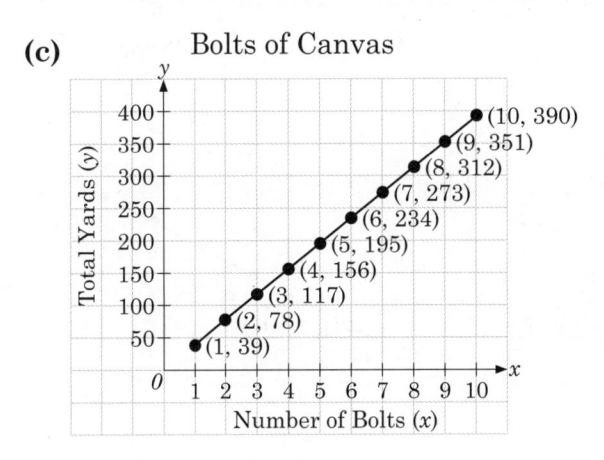

   (c) Bolts of Canvas

   Graph with points (1, 39), (2, 78), (3, 117), (4, 156), (5, 195), (6, 234), (7, 273), (8, 312), (9, 351), (10, 390).

9. **(a)** Complete the table: (4, 14), (5, 17.5)
   $y = 3.5x$
   (b) Complete the table: (15, 22), (30, 37)
   $y = x + 7$
   (c) Complete the table: (25, 24.5), (27.5, 27)
   $y = x - 0.5$

10. **(a)** $y = 3x$
    (b) $t = 2s$
    (c) $q = 50p$

Practice

11. $t = \$11 + \$2.50h$

Time in Hours h	0	0.5	1	1.5	2	2.5	3	3.5
Total Cost in Dollars t	0	13.50	16.00	18.50	21.00	23.50	26.00	28.50

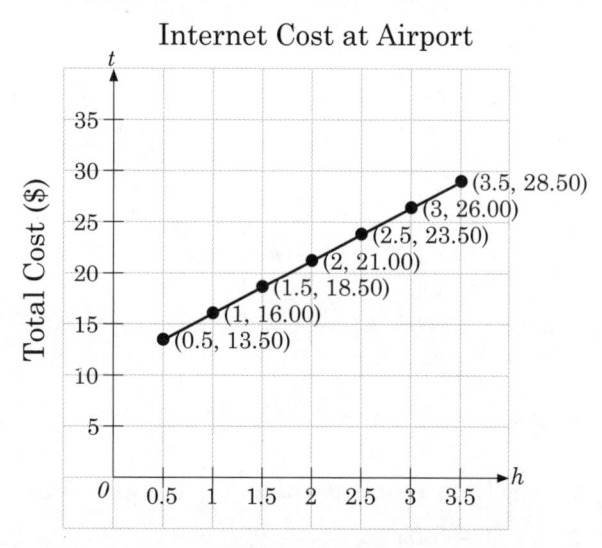

Internet Cost at Airport

Points plotted: (0.5, 13.50), (1, 16.00), (1.5, 18.50), (2, 21.00), (2.5, 23.50), (3, 26.00), (3.5, 28.50)

Axes: Time (hours) vs Total Cost ($)

12. $c = \$8 + \$3.25m$

Number of Miles Biked m	Total Cost in Dollars c
1	11.25
3	17.75
7	30.75
12	47.00
14	53.50
18	66.50
20	73.00

Messenger Service Charges

Points plotted: (1, 11.25), (3, 17.75), (7, 30.75), (12, 47.00), (14, 53.50), (18, 66.50), (20, 73.00)

Axes: Distance Biked (miles) vs Total Cost of Delivery ($)

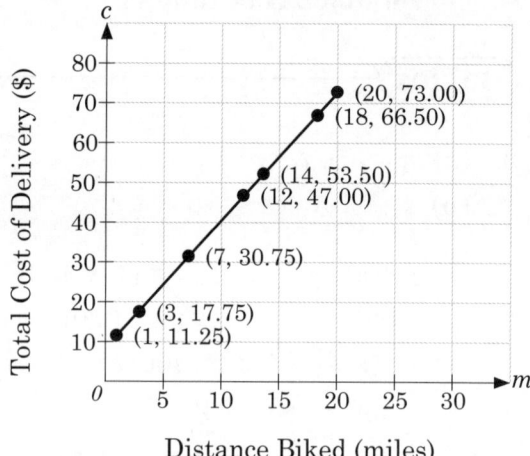

13. $t = \$250 + \$45g$

Rounds of Golf g	Cost in Dollars t	Rounds of Golf g	Cost in Dollars t
1	295	14	880
2	340	15	925
3	385	16	970
4	430	17	1,015
5	475	18	1,060
6	520	19	1,105
7	565	20	1,150
8	610	21	1,195
9	655	22	1,240
10	700	23	1,285
11	745	24	1,330
12	790	25	1,375
13	835		

Cost for Rounds of Golf

Points plotted: (1, 295), (5, 475), (10, 700), (12, 790), (18, 1,060), (21, 1,195), (25, 1,375)

Axes: Rounds of Golf vs Cost ($)

Chapter 10 **COORDINATES AND GRAPHS**

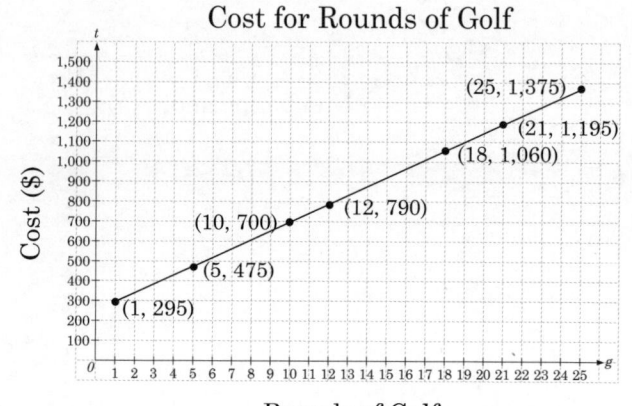

14. (a) Answers will vary.

Example solution:

Square feet f	120	132	144	156	168
Cost in Dollars t	2,640	2,904	3,168	3,432	3,696

(b) $t = \$22f$

(c) Points on graph will vary.

Cost of Patio

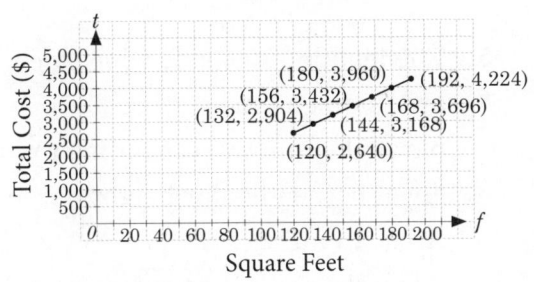

Points shown: (120, 2,640), (132, 2,904), (144, 3,168), (156, 3,432), (168, 3,696), (180, 3,960), (192, 4,224)

x-axis: Square Feet (f); y-axis: Total Cost ($) ($t$)

15. (a)

Number of Hours h	Number of Gallons g
0.5	8
1	16
2	32
4	64
7.5	120
10.5	168

(b) $g = 16h$

(c) Rate a Hose Can Fill a Pool

Points: (0.5, 8), (1, 16), (2, 32), (4, 64), (7.5, 120), (10.5, 168)

x-axis: Number of Hours (h); y-axis: Number of Gallons (g)

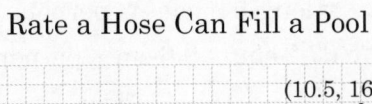

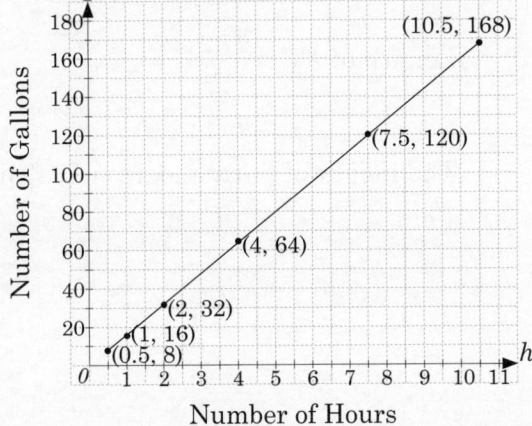

Challenge

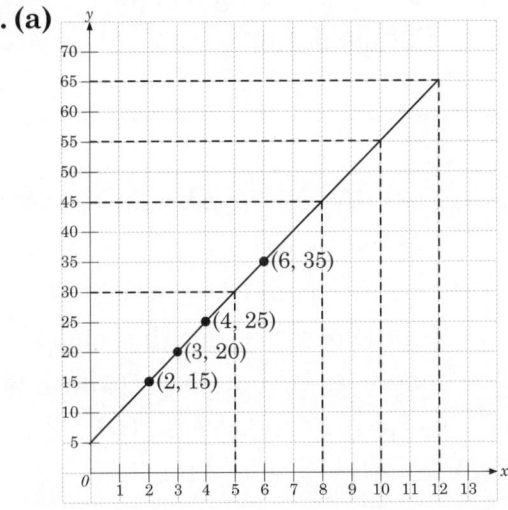

16. (a) Graph with points (2, 15), (3, 20), (4, 25), (6, 35) and line extending through (8, 45), (10, 55), (12, 65).

(b)

x	2	3	4	5	6	8	10	12
y	15	20	25	30	35	45	55	65

(c) $y = 5 + 5x$ or $y = 5x + 5$

(d) $100 = 5 + 5x$

$5x = 95$

$x = 19$

Chapter 10 **COORDINATES AND GRAPHS** 63

Chapter 11: Area of Plane Figures

11.1 Area of Rectangles and Parallelograms

Basics

1. (a) Yes. MN is perpendicular to base CD and to base AB.
   (b) Yes. Although CQ is outside the parallelogram it has the same length as MN and is perpendicular to base AB and to base CD.
   (c) No. AD is a side of the parallelogram. It is neither perpendicular to BC nor CD.
   (d) Yes. OP is perpendicular to base BC and to base AD.

2. The line segment which gives the height of parallelogram $JKLM$ corresponding to base LM is NO.
   Another base and corresponding height for the parallelogram: base KL, height PQ.

3. (a) A base is EF and the corresponding height is GH.
   Area of parallelogram $CDEF$:
   10 cm × 4 cm = 40 cm²
   (b) A base is AB and the corresponding height is EF.
   Area of parallelogram $ABCD$:
   8 cm × 5.8 cm = 46.4 cm²
   (c) A base is DE. The corresponding height, which is outside the parallelogram, is FG.
   Area of parallelogram $CDEF$:
   3 cm × 4 cm = 12 cm²

Practice

4. Area of small parallelogram:
   0.6 m × 0.4 m = 0.24 m²

   Area of large parallelogram:
   0.8 m × 0.4 m = 0.32 m²

   Area of figure:
   0.24 m² + 0.32 m² = 0.56 m²

5. The length of the side which we are finding is the base of the parallelogram corresponding to the given perpendicular height.
   Area of mural
   = length of side × perpendicular distance
   22.8 m² = length of side × 3.8 m
   Length of side = $\frac{22.8 \text{ m}^2}{3.8 \text{ m}}$
   = 6 m
   The length of the longer side is 6 m.

6. The length we are finding is the height of the parallelogram corresponding to the given length of base VW. XV is a height of the parallelogram because it is the perpendicular distance between lines TU and VW.
   Area of parallelogram
   = length of side × perpendicular distance
   67.2 cm² = 9.6 cm × perpendicular distance
   Length of XV = $\frac{67.2 \text{ cm}^2}{9.6 \text{ cm}}$
   = 7 cm
   The height of parallelogram $TUVW$ is 7 cm.

7. Area of billboard:
   26 ft × 8 ft = 208 ft²
   Find $\frac{3}{8}$ of 208 ft².
   $\frac{3}{8} \times 208$ ft² = 78 ft²
   The area of the billboard that is blue is 78 ft².

8. Area of pendant:
   6.2 cm × 10.3 cm = 63.86 cm²
   $\frac{1}{4} + \frac{1}{6} + \frac{1}{12} = \frac{1}{2}$
   $1 - \frac{1}{2} = \frac{1}{2}$
   $\frac{1}{2} \times 63.86$ cm² = 31.93 cm²

   Rounded to the nearest tenth, 31.9 cm² of the pendant is covered in sapphires.

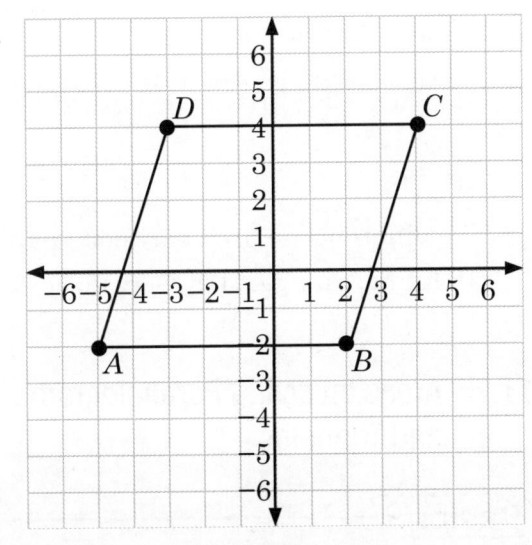

9. Area of Parallelogram ABCD:
   7 units × 6 units = 42 units²

(Challenge)

10. Area of rectangle:
    $\frac{5}{4}$ yd × $\frac{2}{3}$ yd = $\frac{5}{6}$ yd²
    Area of each piece:
    $\frac{1}{2} \times \frac{5}{6}$ yd² = $\frac{5}{12}$ yd²
    The area of each piece is $\frac{5}{12}$ yd².

11. Area of smaller parallelogram:
    13.5 cm²
    Base of smaller parallelogram:
    13.5 cm² ÷ 3 cm = 4.5 cm
    Base of larger parallelogram:
    2 × 4.5 cm = 9 cm

    Height of larger parallelogram:
    2 units ⟶ 3 cm
    1 unit ⟶ 1.5 cm
    3 units ⟶ 3 × 1.5 cm = 4.5 cm

    Area of larger parallelogram:
    9 cm × 4.5 cm = 40.5 cm²
    Total area of two parallelograms:
    13.5 cm² + 40.5 cm² = 54 cm²

Area of Triangles

11.2A Finding Area of a Triangle

(Basics)

1. (a) Yes. OL is perpendicular to base MN and L is the vertex opposite MN.
   (b) No. Although PQ is perpendicular to base LM, it does not extend to the opposite vertex.
   (c) No. MN is not perpendicular to any base.

2. (a) Answers will vary.
       Possible answers:
       Base MN, height OP
       Base NO, height MQ
       Base MO, height NR
   (b) Answers will vary.
       Possible answers:
       Base ST, height TU
       Base TU, height ST
       Base SU, height TV

3. Triangle *DEF* is a right triangle. If the base is *DF*, the corresponding height is *DE*. If the base is *DE*, the corresponding height is *DF*.
   Area of triangle *DEF*:
   $\frac{1}{2} \times 8 \text{ ft} \times 15 \text{ ft} = 60 \text{ ft}^2$

4. Area of triangle *ABC*:
   $= \frac{1}{2} \times 19\frac{1}{2} \text{ cm} \times 4 \text{ cm} = 39 \text{ cm}^2$

   Area of triangle *ABD*:
   $= \frac{1}{2} \times 3 \text{ cm} \times 4 \text{ cm} = 6 \text{ cm}^2$

   Area of triangle *DBC*:
   $= \frac{1}{2} \times 16\frac{1}{2} \text{ cm} \times 4 \text{ cm} = 33 \text{ cm}^2$

Practice

5. (a) Area of triangle *QRS*:
   $0.5 \times 2.5 \text{ cm} \times 4.5 \text{ cm} = 5.625 \text{ cm}^2$
   The area of triangle *QRS* rounded to one decimal place is 5.6 cm².

   (b) The base is *XY* and the height is *WZ*.
   WZ = 3 in
   XY = *ZY* − *XZ* = (6.1 − 3.2) in = 2.9 in
   $\frac{1}{2} \times 3 \text{ in} \times 2.9 \text{ in} = 4.35 \text{ in}^2$

   The area of triangle *WXY* rounded to one decimal place is 4.4 in².

6. We can consider the opposite edge as the base and the perpendicular distance from the top vertex to the opposite edge as the height.
   $140 \text{ ft}^2 = \frac{1}{2} \times b \times 8 \text{ ft}$
   $\frac{2(140 \text{ ft}^2)}{8 \text{ ft}} = b$
   35 ft = *b*
   The length of this edge is 35 ft.

Challenge

7. Area of *GHK*:
   $\frac{1}{2} \times 2 \text{ in} \times 2\frac{3}{4} \text{ in}$
   $= 1 \text{ in} \times 2\frac{3}{4} \text{ in}$
   $= 2\frac{3}{4} \text{ in}^2$

   Area of *HIK*:
   $\frac{1}{2} \times 1\frac{3}{4} \text{ in} \times 2\frac{3}{4} \text{ in}$
   $= \frac{1}{2} \times \frac{7}{4} \text{ in} \times \frac{11}{4} \text{ in}$
   $= \frac{77}{32} \text{ in}^2$
   $= 2\frac{13}{32} \text{ in}^2$

   Difference between the two areas:
   $2\frac{3}{4} \text{ in}^2 - 2\frac{13}{32} \text{ in}^2$
   $= 2\frac{24}{32} \text{ in}^2 - 2\frac{13}{32} \text{ in}^2$
   $= \frac{11}{32} \text{ in}^2$

   The difference in area between triangle *GHK* and triangle *HIK* is $\frac{11}{32}$ in².

11.2B Areas Involving Parallelograms and Triangles

Basics

8. The area of triangle *FGH* = $\frac{1}{2}$ × the area of parallelogram *FGHI*.
   $\frac{1}{2} \times 38.25 \text{ cm}^2 = 19.125 \text{ cm}^2$
   The area of triangle *FGH* rounded to two decimal places is 19.13 cm².

9. Area of fabric:
   3 ft × 4 ft = 12 ft²

   Area of banner:
   $\frac{1}{2} \times 2\frac{1}{2}$ ft × 4 ft = $\frac{1}{2} \times \frac{5}{2}$ ft × 4 ft
   = 5 ft²

   Fabric left over:
   12 ft² − 5 ft² = 7 ft²

   There will be 7 ft² of fabric left over.

10. Area:
    $\frac{1}{2}$ × 15 m × 6 m = 45 m²

11. The length of x:
    96 cm² ÷ 16 cm = 6 cm

Practice

12. To find the base of triangle *NPQ*,
    4 units ⟶ 2 ft
    1 unit ⟶ $\frac{1}{4}$ × 2 ft = $\frac{1}{2}$ ft
    3 units ⟶ 3 × $\frac{1}{2}$ ft = $\frac{3}{2}$ ft

    Area of triangle *NPQ* = $\frac{1}{2} \times \frac{3}{2}$ ft × 2 ft
    = $\frac{3}{2}$ ft²
    = $1\frac{1}{2}$ ft²

13. Area of parallelogram *MNOP*:
    4 cm × 4 cm = 16 cm²

Challenge

14. Area of flag: 7 ft × 3 ft = 21 ft²

    Area of figure A (a parallelogram): bh

    Base of figure A:
    7 ft − $2\frac{1}{4}$ ft − 3 ft = $1\frac{3}{4}$ ft

    Height of figure A:
    3 ft − $1\frac{1}{4}$ ft = $1\frac{3}{4}$ ft

    Area of figure A:
    $1\frac{3}{4}$ ft × $1\frac{3}{4}$ ft = $\frac{7}{4}$ ft × $\frac{7}{4}$ ft
    = $\frac{49}{16}$ ft² = $3\frac{1}{16}$ ft²

    Area of figure B (a triangle): $\frac{1}{2}bh$
    Base of figure B: 3 ft
    Height of figure B: $1\frac{3}{4}$ ft
    Area of figure B:
    $\frac{1}{2}$ × 3 ft × $1\frac{3}{4}$ ft = $\frac{1}{2}$ × 3 ft × $\frac{7}{4}$ ft
    = $\frac{21}{8}$ ft² = $2\frac{5}{8}$ ft²

    Combined area of figures A and B:
    $3\frac{1}{16}$ ft² + $2\frac{5}{8}$ ft² = $5\frac{11}{16}$ ft²

    The combined area of figures A and B is $5\frac{11}{16}$ ft².

11.3 Area of Trapezoids

Basics

1. (a) No. *FG* is not a height. It is a base.
   (b) No. *GH* is not a height, as it is not perpendicular to the bases.
   (c) Yes. *GJ* is a height as it is perpendicular to *FG* and *HI*.
   (d) Yes. Although *IK* is outside the trapezoid, it is a perpendicular line to bases *FG* and *HI*. Thus, *IK* is a height of trapezoid *FGHI* between the bases.

2. **(a)** The bases are the parallel sides, *HI* and *JK*. The height is *LM*.
   Area of trapezoid *HIJK*:
   $\frac{1}{2} \times (6 \text{ in} + 13 \text{ in}) \times 7 \text{ in}$
   $= \frac{1}{2} \times 19 \text{ in} \times 7 \text{ in}$
   $= 66.5 \text{ in}^2$

   (b) The bases are the parallel sides, *RU* and *ST*. *TU* is the height.
   Area of trapezoid *RSTU*:
   $\frac{1}{2} \times (9 \text{ m} + 5 \text{ m}) \times 38 \text{ m}$
   $= \frac{1}{2} \times 14 \text{ m} \times 38 \text{ m}$
   $= 266 \text{ m}^2$

   (c) The bases are the parallel sides, *VW* and *XY*. The height is *WZ*.
   Area of trapezoid *VWXY*:
   $\frac{1}{2} \times (4.1 \text{ m} + 2 \text{ m}) \times 8.6 \text{ m}$
   $= 26.23 \text{ m}^2$

Practice

3. Area of one trapezoid:
   $\frac{1}{2} \times (5 \text{ cm} + 2\frac{1}{2} \text{ cm}) \times 4 \text{ cm}$
   $= \frac{1}{2} \times 7\frac{1}{2} \text{ cm} \times 4 \text{ cm}$
   $= \frac{1}{2} \times \frac{15}{2} \text{ cm} \times 4 \text{ cm}$
   $= 15 \text{ cm}^2$
   Area of hexagon: $2 \times 15 \text{ cm}^2 = 30 \text{ cm}^2$

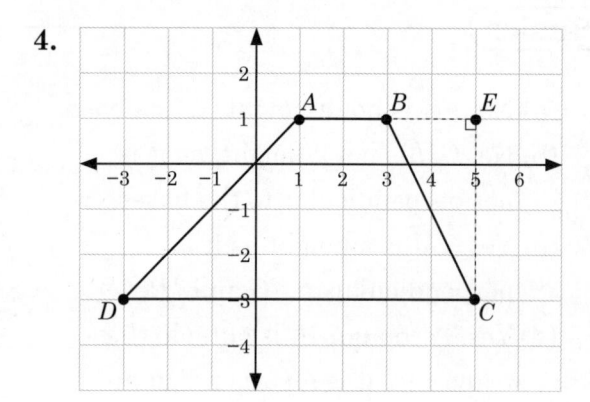

4.

The bases are the parallel sides, *AB* and *CD*.
The height is CE since it is a perpendicular line between the bases.
Let the length of CD be x cm.
Area of trapezoid *ABCD* = 20 cm²
Using the formula,
Area $= \frac{1}{2} \times (b1 + b2) \times h$:
$\frac{1}{2} \times (2 \text{ cm} + 8 \text{ cm}) \times 4 \text{ cm}$
$= \frac{1}{2} \times 10 \text{ cm} \times 4 \text{ cm} = 20 \text{ cm}^2$
$20 \text{ cm}^2 = \frac{1}{2} \times (2 \text{ cm} + x) \times 4 \text{ cm}$
$20 \text{ cm}^2 = 2 \text{ cm} \times (2 \text{ cm} + x)$
$\frac{20}{2} \text{ cm} = \frac{2}{2} \text{ cm} \times (2 \text{ cm} + x)$
$10 \text{ cm} = 2 \text{ cm} + x$
$10 \text{ cm} - 2 \text{ cm} = 2 \text{ cm} - 2 \text{ cm} + x$
$x = 8 \text{ cm}$
The length of CD is 8 cm, therefore the coordinates of D is (−3, −3).

Challenge

5. Answers will vary. A correct answer will show a trapezoid with an area of 18.75 cm².
   The bases are parallel sides AB and CD.
   The height is 6 units, or 3 cm.
   Area of trapezoid ABCD:
   $\frac{1}{2} \times 12.5 \text{ cm} \times 3 \text{ cm}$
   $= 18.75 \text{ cm}^2$
   Example solution:
   Length of AB is 10 cm.
   Length of CD is 15 cm.
   Height is 1.5 cm.

Chapter 12: Volume and Surface Area of Solids

Volume of Rectangular Prisms
12.1A Cubes and Cuboids

Basics

1. $\frac{3}{4}$ in $\times \frac{3}{4}$ in $\times \frac{3}{4}$ in

   $= (\frac{3}{4} \times \frac{3}{4} \times \frac{3}{4})$ in^3

   $= \frac{27}{64}$ in^3

   The volume is $\frac{27}{64}$ in^3.

2. $1\frac{1}{2}$ ft $\times 7\frac{1}{2}$ ft $\times 1\frac{1}{2}$ ft

   $= (1\frac{1}{2} \times 7\frac{1}{2} \times 1\frac{1}{2})$ ft^3

   $= (\frac{3 \times 15 \times 3}{2 \times 2 \times 2})$ ft^3

   $= \frac{135}{8}$ ft^3

   $= 16\frac{7}{8}$ ft^3

   The volume is $16\frac{7}{8}$ ft^3.

3. (a) 7.2 cm × 3.5 cm × 4 cm = 100.8 cm^3

   The volume is 100.8 cm^3.

   (b) 15 in × $4\frac{4}{5}$ in × $5\frac{1}{2}$ in = 396 in^3

   The volume is 396 in^3.

4. $16\frac{2}{3}$ ft × 42 ft^2

   $= \frac{50}{\cancel{3}_1} \times \frac{\cancel{42}^{14}}{1}$

   $= 700$ ft^3

   The volume of the rectangular prism is 700 ft^3.

5. 12 cm × 5.7 cm × w cm = 410.4 cm^3

   $w = 410.4$ cm^3 ÷ (12 cm × 5.7 cm)

   $w = \frac{410.4 \text{ cm}^3}{68.4 \text{ cm}^2} = 6$ cm

   The width of the rectangular prism is 6 cm.

Practice

6. Height of rectangular prism A:
   72 cm^3 ÷ 9 cm^2 = 8 cm

   Height of rectangular prism B:
   8 cm ÷ 4 = 2 cm

   Base of rectangular prism:
   72 cm^3 ÷ 2 cm = 36 cm^2

   Answers will vary. The area of the base of rectangular prism B must be 36 cm^2.

   Possible solutions:
   1 cm by 36 cm by 2 cm
   2 cm by 18 cm by 2 cm
   3 cm by 12 cm by 2 cm
   6 cm by 6 cm by 2 cm
   9 cm by 4 cm by 2 cm

7. If the figure is divided as shown:

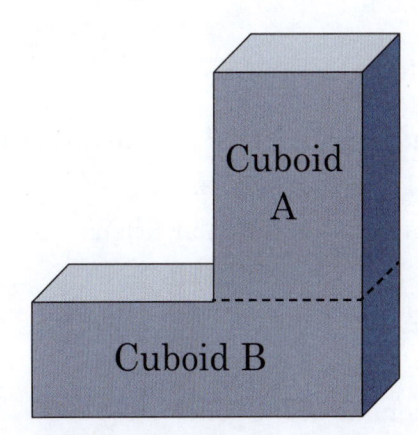

   Volume of cuboid A:
   3.4 ft × 6.5 ft × 4 ft = 88.4 ft^3

   Volume of cuboid B:
   9 ft × 4 ft × 4 ft = 144 ft^3

   Volume of solid:
   88.4 ft^3 + 144 ft^3 = 232.4 ft^3

   The volume is 232.4 ft^3.

8. 7.75 L + 3.25 L = 11 L = 11,000 cm³

   25 cm × 22 cm = 550 cm²

   $\frac{11,000 \text{ cm}^3}{550 \text{ cm}^2}$ = 20 cm

   The height is 20 cm.

9. Rectangular prism with no pieces removed:

   25 in × 20 in × 4.5 in = 2,250 in³

   Smallest missing rectangular prism:

   3 in × 8 in × 4.5 in = 108 in³

   Medium missing rectangular prism:

   5 in × 11 in × 4.5 in = 247.5 in³

   Volume of the solid:

   2,250 in³ − 108 in³ − 247.5 in³
   = 1,894.5 in³

   The volume is 1,894.5 in³.

10. Length of one diagonal rectangular prism:

    6 in + 3 in + 6 in = 15 in

    One diagonal rectangular prism:

    15 in × 3 in × 3 in = 135 in³

    Two smaller rectangular prisms:

    2(6 in × 3 in × 3 in) = 108 in³

    Total volume of the solid:

    135 in³ + 108 in³ = 243 in³

    The volume is 243 in³.

Challenge

11. **(a)** Volume:

    8 ft × 2 ft × 3 ft = 48 ft³

    The volume is 48 ft³.

    (b) Answers will vary. Volume must be 48 ft³. Examples:

    4 ft × 4 ft × 3 ft, 16 ft × 3 ft × 1 ft,

    12 ft × 4 ft × 1 ft, 6 ft × 8 ft × 1 ft,

    24 ft × 2 ft × 1 ft

12. Total volume:

    5 cm × 12 cm × 8 cm = 480 cm³

    Volume of hole:

    2.5 cm × 2.5 cm × 12 cm = 75 cm³

    Volume of drilled block:

    480 cm³ − 75 cm³ = 405 cm³

    The volume is 405 cm³.

12.1B Volume of Liquids

Basics

13. Volume of tank:

    24 cm × 12 cm × 12 cm = 3,456 cm³

    Volume of water in tank:

    $\frac{7}{8}$ × 3,456 cm³ = 3,024 cm³

    The volume is 3,024 cm³.

14. Volume of cubical container:

    4 cm × 4 cm × 4 cm
    = 64 cm³

    Volume of rectangular container:

    12 cm × $2\frac{2}{3}$ cm × 4 cm
    = 128 cm³

    Fraction of rectangular container that will be filled = $\frac{64 \text{ cm}^3}{128 \text{ cm}^3} = \frac{1}{2}$

    $\frac{1}{2}$ of the rectangular container will be filled.

15. 45 L × 2 = 90 L = 90,000 cm³

    90,000 cm³ ÷ 450 cm² = 200 cm

    The height is 200 cm.

16. 10.5 cm × 13 cm × 15.5 cm
    = 2,115.75 cm³

    2,115.75 cm³ − 1,500 cm³ = 615.75 cm³

    615.75 cm³ of space is not filled with water.

Practice

17. Volume of water remaining in rectangular container:
$\frac{2}{3} \times 15$ cm $\times 5$ cm $\times 8$ cm
$= 400$ cm³
Answers will vary. The volume of the smaller container must be 400 cm³.
Example answers: 10 cm × 10 cm × 4 cm, 5 cm × 4 cm × 20 cm

18. Volume of the water in the tank when full:
60 cm × 40 cm × 50 cm = 120,000 cm³
$\frac{3}{4} \times 120{,}000$ cm³ = 90,000 cm³
90,000 cm³ = 90 L

$\frac{90 \text{ L}}{10 \text{ L}} = 9$

It will take 9 minutes to empty the tank.

19. 90 cm × 30 cm × 45 cm = 121,500 cm³
121,500 cm³ ÷ 18,000 cm³ = 6.75
It will take 6.75 minutes.

20. 42 cm × 25 cm = 1,050 cm²
$\frac{64 \text{ cm}^3}{128 \text{ cm}^3} = 5.5$ cm

5.5 cm + 19.5 cm = 25 cm
The height is 25 cm.

21. 150 L = 150,000 cm³
Base of tank A:
50 cm × 45 cm = 2,250 cm²
Volume of tank A:
2,250 cm² × 70 cm = 157,500 cm³
= 157 L 500 mL
Base of tank B:
55 cm × 60 cm = 3,300 cm²
Volume of tank B:
3,300 cm² × 50 cm = 165,000 cm³
= 165 L
Sydney should buy tank A. It has more than the minimum volume she wants and will take up less room on her desk.

Challenge

22. Initial water level in the pool:
$\frac{3}{5} \times 6$ ft $= 3\frac{3}{5}$ ft
Volume of water increase per minute:
1 ft³
Volume of water decrease per minute:
0.25 ft³
1 ft³ − 0.25 ft³ = 0.75 ft³ increase per minute
Total increase in volume of water after 30 minutes:
30 × 0.75 ft³ = 22.5 ft³
Increase in water level =
$\frac{\text{total increase in volume}}{\text{length} \times \text{width of pool}}$

$= \frac{22.5 \text{ ft}^3}{12 \text{ ft} \times 15 \text{ ft}^3}$

$= \frac{1}{8}$ ft

New water level after 30 minutes:
$3\frac{3}{5}$ ft $+ \frac{1}{8}$ ft $= 3\frac{29}{40}$ ft $= 3.725$ ft

The new water level after 30 minutes is 3.725 ft.

23. Method 1

Volume of water in the tank before the cube is lowered:
16 cm × 24 cm × 20 cm = 7,680 cm³

Volume of the cube:
12 cm × 12 cm × 12 cm = 1,728 cm³

Volume of the water and volume of the cube:
7,680 cm³ + 1,728 cm³ = 9,408 cm³

New height of the water:
9,408 cm³ ÷ (16 cm × 24 cm) = 24.5 cm

Difference between the new height of the water and the original height of the water: 24.5 cm − 20 cm = 4.5 cm

Method 2

Volume of the cube:
12 cm × 12 cm × 12 cm = 1,728 cm³

Height of this volume in the tank:
1,728 cm³ ÷ (16 cm × 24 cm) = 4.5 cm

The difference between the two water heights is 4.5 cm.

Surface Area of Prisms
12.2A Surface Area of Rectangular Prisms

Basics

1. Area of each face of the cube:
$\frac{54 \text{ ft}^2}{6} = 9 \text{ ft}^2$

The face of a cube is a square.
3 × 3 = 9, so each face measures 3 ft.
If the area of a face of the cube is 9 ft², then each side measures 3 ft.

Volume of the cube:
3 ft × 3 ft × 3 ft = 27 ft³.

The volume of the cube is 27 ft³.

2. (a) $(lw + lh + wh) \times 2$
$= (2.5 \times 3.5 + 2.5 \times 4 + 3.5 \times 4) \text{ cm}^2 \times 2$
$= (8.75 + 10 + 14) \text{ cm}^2 \times 2$
$= 32.75 \text{ cm}^2 \times 2$
$= 65.5 \text{ cm}^2$

The surface area is 65.5 cm².

(b) $(lw + lh + wh) \times 2$
$= (25\frac{1}{2} \times 5\frac{1}{2} + 25\frac{1}{2} \times 6 + 5\frac{1}{2} \times 6) \text{ ft}^2 \times 2$
$= (140\frac{1}{4} + 153 + 33) \text{ ft}^2 \times 2$
$= 326\frac{1}{4} \text{ ft}^2 \times 2$
$= 652\frac{1}{2} \text{ ft}^2$

The surface area is $652\frac{1}{2}$ ft².

3. $s^2 \times 6$
$= (3.5 \times 3.5) \text{ cm}^2 \times 6$
$= 12.25 \text{ cm}^2 \times 6$
$= 73.5 \text{ cm}^2$

The surface area of the cube is 73.5 cm².

Practice

4.

Prism	Volume	Surface Area
a	2m × 1.5 m × 4 m = 12 m³	(2 × 1.5 + 2 × 4 + 1.5 × 4) m² × 2 = (3 + 8 + 6) m² × 2 = 17 m² × 2 = 34 m²
b	4.5 ft × 3.5 ft × 5 ft = 78.75 ft³	(4.5 × 3.5 + 4.5 × 5 + 3.5 × 5) ft² × 2 = (15.75 + 22.5 + 17.5) ft² × 2 = 55.75 ft² × 2 = 111.5 ft²
c	55 cm × 35 cm × 60 cm = 115,500 cm³	(55 × 35 + 55 × 60 + 35 × 60) cm² × 2 = (1,925 + 3,300 + 2,100) cm² × 2 = 7,325 cm² × 2 = 14,650 cm²
d	1.5 in × 2 in × 1.75 in = 5.25 in³	(1.5 × 2 + 1.5 × 1.75 + 2 × 1.75) in² × 2 = (3 + 2.625 + 3.5) in² × 2 = 9.125 in² × 2 = 18.25 in²

5. (a) Volume: $\frac{9}{2}$ cm × $\frac{9}{2}$ cm × $\frac{9}{2}$ cm
= $\frac{729}{8}$ cm³ = $91\frac{1}{8}$ cm³

The volume is $91\frac{1}{8}$ cm³.

Surface Area:
$6(\frac{9}{2}$ cm × $\frac{9}{2}$ cm$) = 6 \times \frac{81}{4}$ cm²
= $121\frac{1}{2}$ cm²

The surface area is $121\frac{1}{2}$ cm².

(b) Volume: 7 m × 7 m × 7 m = 343 m³

The volume is 343 m³.

Surface Area:
6(7 m × 7 m) = 6 × 49 m² = 294 m²

The surface area is 294 m².

6. Surface area of Box A:
(10 × 13 + 10 × 3.5 + 3.5 × 13) cm² × 2
= (130 + 35 + 45.5) cm² × 2
= 210.5 cm² × 2
= 421 cm²

Surface area of Box B:
(10 × 8 + 10 × 5 + 8 × 5) cm² × 2
= (80 + 50 + 40) cm² × 2
= 170 cm² × 2
= 340 cm²

Difference between the surface area of the two boxes:
(421 − 340) cm² = 81 cm²

The difference between the surface area of the two boxes is 81 cm².

Challenge

7. (a) Volume of solid:

Larger rectangular prism:
12 ft × 12 ft × 1.5 ft = 216 ft³

Front rectangular prism:
1.5 ft × 1.5 ft × 3 ft = 6.75 ft³

Back rectangular prism:
1.5 ft × 1.5 ft × 5 ft = 11.25 ft³

Total volume:
216 ft³ + 6.75 ft³ + 11.25 ft³
= 234 ft³

The volume of the solid is 234 ft³.

(b) Surface area of solid:

Larger rectangular prism:
Two large horizontal surfaces:
2(12 ft × 12 ft) = 288 ft²

Surface area of the sides of the small rectangular prisms the rest on the large rectangular prism:
2(1.5 ft × 1.5 ft) = 4.5 ft²
288 ft² − 4.5 ft² = 283.5 ft²

Four smaller rectangular surfaces:
4(12 ft × 1.5 ft) = 72 ft²

Total surface area of the larger rectangular prism:
283.5 ft² + 72 ft² = 355.5 ft²

Surface area of front rectangular prism:
4(1.5 ft × 3 ft) + (1.5 ft × 1.5 ft)
= 18 ft² + 2.25 ft² = 20.25 ft²

Surface area of back rectangular prism:
4(1.5 ft × 5 ft) + (1.5 ft × 1.5 ft)
= 30 ft² + 2.25 ft² = 32.25 ft²

Total surface area of the solid:
355.5 ft² + 20.25 ft² + 32.25 ft² = 408 ft²

The total surface area of the solid is 408 ft².

8. (c) and (e) are not nets of cubes.

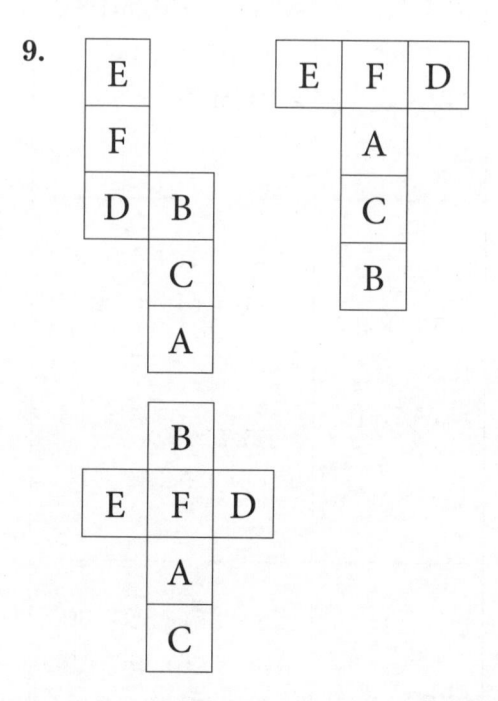

9.

	E					E	F	D
	F						A	
D	B						C	
	C						B	
	A							

	B		
E	F	D	
	A		
	C		

12.2B Surface Area of Triangular Prisms

Basics

10. Area of triangular bases:
$\frac{1}{2} \times 24$ m $\times 5$ m $\times 2 = 120$ m²

Area of bottom face:
24 m × 30 m = 720 m²

Area of rectangular sides:
30 m × 13 m × 2 = 780 m²

Total surface area of storage unit:
120 m² + 720 m² + 780 m² = 1,620 m²

The surface area is 1,620 m².

11. Area of triangular sides:
$2 \times (\frac{1}{2} \times 3 \text{ cm} \times 4 \text{ cm}) = 12 \text{ cm}^2$

Area of base:
6 cm × 3 cm = 18 cm²

Area of front rectangle:
6 cm × 5 cm = 30 cm²

Area of back rectangle:
6 cm × 4 cm = 24 cm²

Total surface area:
12 cm² + 18 cm² + 30 cm² + 24 cm²
= 84 cm²

The surface area is 84 cm².

Practice

12. Area of rectangular faces:
2 × (10 cm × 5 cm) = 100 cm²

Area of rectangular base:
10 cm × 8 cm = 80 cm²

Total surface area of rectangular faces:
100 cm² + 80 cm² = 180 cm²

Cost of fabric to produce one box:
$0.02 × 180 = $3.60

The cost for the fabric to produce one such box is $3.60.

13. Area of triangular bases:
$\frac{1}{2} \times 24 \text{ m} \times 5 \text{ m} \times 2 = 120 \text{ m}^2$

Area of bottom face:
$24 \text{ m} \times 22\frac{1}{2} \text{ m} = 540 \text{ m}^2$

Area of rectangular faces:
$22\frac{1}{2} \text{ m} \times 13 \text{ m} \times 2 = 585 \text{ m}^2$

Total surface area of storage unit:
120 m² + 540 m² + 585 m² = 1,245 m²

The surface area would be 1,245 m².

12.2C Metric Conversions and Volume

Basics

14. (a) 4.5 m = 450 cm
3.5 m = 350 cm
1.9 m = 190 cm

Volume in m³:
4.5 m × 3.5 m × 1.9 m = 29.925 m³

Volume in cm³:
450 cm × 350 cm × 190 cm
= 29,925,000 cm³

The volume is 29.925 m³ or 29,925,000 cm³.

(b) Surface area in m²: 2(4.5 m × 1.9 m + 3.5 m × 4.5 m + 1.9 m × 3.5 m)
= 2(8.55 m² + 15.75 m² + 6.65 m²)
= 2(30.95 m²)
= 61.9 m²

Surface area in cm²:
2(450 cm × 190 cm + 350 cm × 450 cm + 190 cm × 350 cm)
= 2(85,500 cm² + 157,500 cm² + 66,500 cm²)
= 2(309,500 cm²)
= 619,000 cm²

The surface area is 61.9 m² or 619,000 cm².

Practice

15. (a) 18 in = 1.5 ft

Volume of sandbox:
4 ft × 3 ft × 1.5 ft = 18 ft³

The sandbox will hold 18 ft³ of sand when completely full.

(b) 4 ft = 48 in
3 ft = 36 in
18 in − 3 in = 15 in

Volume of total sand needed:
48 in × 36 in × 15 in = 25,920 in³

Colored sand: 25% × 25,920 in³
= $\frac{25}{100}$ × 25,920 in³ = 6,480 in³

They used 6,480 in³ of colored sand.

Natural sand:
25,920 in³ − 6,480 in³ = 19,440 in³

They used 19,440 in³ of natural sand.

16. Tarp A: 120 cm²
Tarp B: 1,350 cm²
Tarp C: 23.45 cm²
Tarp D: 134,500 cm²
The area of the tarps from largest to smallest is Tarp D, Tarp B, Tarp A, and Tarp C.

17. Area of wall opposite door:
28 ft × 9 ft = 252 ft²

Area of one of the adjacent walls:
32 ft × 9 ft = 288 ft²

Area of ceiling:
28 ft × 32 ft = 896 ft²

Total area to be painted:
(252 + 288 + 288 + 896) ft² = 1,724 ft²

1,724 ft² ÷ 300 ft² is about 5.75 gallons.

Cost of paint: 6 gallons × $18 = $108

The paint for this project will cost $108.

Chapter 13: Displaying and Comparing Data

Statistical Variability
13.1AB Statistical Questions & Measures of Center

Basics

1. **(a)** This is not a statistical question. There is only one answer.
   (b) This is a statistical question because the answers may vary.
   (c) This is a statistical question because it would be answered by collecting data on the prices, and variability in the prices is expected.

2. Mean high temperature:
   $(116 + 112 + 110 + 108 + 108 + 109 + 112) \div 7$
   $= \frac{775}{7} = 110.7$ degrees Fahrenheit
   The mean is 111 degrees Fahrenheit, when rounded to the nearest degree.

3. Total score for five games:
   $5 \times 98 = 490$
   Total score for 4 games:
   $104 + 99 + 95 + 88 = 386$
   Score in fifth game:
   $490 - 386 = 104$
   The team scored 104 points in the fifth game.

4. Total of six numbers: $184.2 \times 6 = 1{,}105.2$
   Total of five numbers: $178 \times 5 = 890$
   The sixth number: $1{,}105.2 - 890 = 215.2$
   The sixth number is 215.2

5. Sum of three numbers:
   $3 \times 20.1 = 60.3$
   $d + 2d + d + 2d = 6d$
   $6d = 60.3$
   $d = \frac{60.3}{6} = 10.05$
   $d = 10.05$
   $e = 2 \times 10.05 = 20.1$
   $20.1 - 10.05 = 10.05$
   The difference between d and e is 10.05.

6. **(a)** Mean weight of nine packages:
   $(46 + 29 + 8 + 78 + 40 + 37 + 39 + 66 + 80)$ lb $\div 9$
   $= \frac{423}{9}$ lb
   $= 47$ lb
   The mean weight is 47 pounds.
   (b) Weight, in pounds, arranged in ascending order:
   8, 29, 37, 39, ⓐ40, 46, 66, 78, 80
   The median weight of the nine packages is 40 pounds.

Practice

7. **(a)** Mean weekly sales (in dollars):
   $(2{,}000 + 3{,}200 + 2{,}800 + 5{,}000 + 1{,}800 + 3{,}600 + 3{,}000 + 4{,}800 + 2{,}500 + 2{,}800 + 1{,}000 + 3{,}000 + 1{,}200 + 2{,}500 + 1{,}800 + 1{,}600) \div 16$
   $= \frac{42{,}600}{16} = 2{,}662.50$
   The mean was $2,662.50.

(b) 1,000 1,200 1,600 1,800 1,800
2,000 2,500 (2,500 2,800) 2,800
3,000 3,000 3,200 3,600 4,800
5,000

2,500 + 2,800 = 5,300

$\frac{5,300}{2} = 2,650$

The median was $2,650.

(c) Max has the lowest total commissions and including him moves the average down. Therefore the mean sales average would increase if Max had not worked in April.

8. Answers will vary. The sum of the six numbers must be 381 and the average of the two numbers in the middle must be 58.
   Example answer: 46, 47, 54, 62, 80, 92

9. **(a)** Total of three numbers: 3 × 4 = 12
   The three numbers are 2, 2, 8.
   (b) Total of three numbers: 3 × 8 = 24
   The three numbers are 7, 7, 10.
   (c) Total of three numbers: 3 × 25 = 75
   Answers will vary.
   Examples: 45, 20, 10; 40, 30, 5
   (d) Sum of four numbers: 4 × 21 = 84
   Answers will vary.
   Examples: 4, 18, 22, 40 or 4, 16, 24, 40

Starting with the last statement, "the least integer is $\frac{1}{10}$ the greatest integer," we can use guess and check to find those two integers. Since the sum of the four numbers is 84, the median is 20, and there are four integers, our choices are limited. If the integers were less than or greater than 4 and 40, the total could not be found using four integers and having a median of 20. Once the least and greatest integers are identified, we can find two other integers with a sum of 40 and a median of 20. Some of these pairs of integers could be 14 and 26, 12 and 28, 10 and 30, etc.

10. Mean:
$(\frac{3}{4} + \frac{2}{3} + \frac{1}{4} + \frac{1}{3} + \frac{1}{8} + \frac{1}{2}) \div 6$
$= (\frac{18}{24} + \frac{16}{24} + \frac{6}{24} + \frac{8}{24} + \frac{3}{24} + \frac{12}{24}) \div 6$
$= \frac{63}{24} \div 6$
$= \frac{63}{24} \times \frac{1}{6}$
$= \frac{7}{16}$

Mean: $\frac{7}{16}$

Median:
$\frac{1}{8}, \frac{1}{4}, \frac{1}{3}, \frac{1}{2}, \frac{2}{3}, \frac{3}{4}$
$(\frac{1}{3} + \frac{1}{2}) \div 2$
$= \frac{5}{6} \div 2$
$= \frac{5}{6} \times \frac{1}{2}$
$= \frac{5}{12}$

Median: $\frac{5}{12}$

Mode: No number occurs more often than any other. Thus, there is no mode.

Challenge

11. (a) Total number of times at bat:
495 + 540 + 536 = 1,571

Total hits: 184 + 192 + 173 = 549

Mean batting average: $\frac{549}{1,571} = 0.349$

Babe Ruth's mean batting average for the three-year period was 0.349.

(b) Total number of times at bat:
638 + 640 + 590 = 1,868

Total hits: 200 + 216 + 204 = 620

Mean batting average: $\frac{620}{1,868} = 0.332$

Jose Altuve's mean batting average for the three-year period was 0.332.

12. (a) Mean score:
(20 + 85 + 90 + 82 + 94 + 88 + 82 + 86 + 95) ÷ 9 = $\frac{722}{9}$
= 80.2 (rounded to the nearest tenth)
The student's mean score, rounded to the nearest tenth, was 80.2.

(b) 20, 82, 82, 85, 86, 88, 90, 94, 95
The student's median score was 86.

(c) The median score is more representative. The mean is much lower than the median because 20 is an outlier.

(d) Total score without the 20:
722 − 20 = 702

$\frac{702}{8} = 87.8$

The students mean score, rounded to the nearest tenth, would be 87.8.

Displaying Numerical Data
13.2A Dot Plots

Basics

1. (a) To find the sum of the heights, convert from feet to inches to avoid fractions.
1 ft = 12 in
Mean:
(78 + 69 + 81 + 80 + 82 + 78 + 75 + 79 + 83 + 76 + 78 + 80 + 77 + 70 + 80 + 79 + 78 + 77 + 79 + 81) in ÷ 20
= 1,560 in ÷ 20
= 78 in

To convert back to feet:
78 in ÷ 12 = 6.5
The mean height is 6.5 ft.

(b)

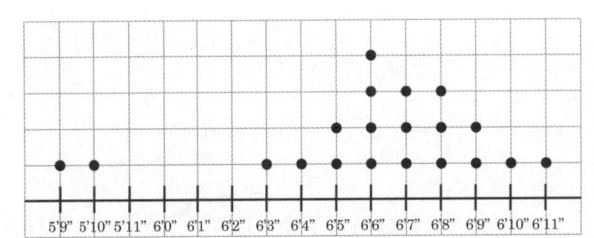

[Dot plot showing heights from 5'9" to 6'11"]

(c) Answers will vary. Answers should include most of the below information.

From the dot plot, we observe the following:
The data varies from 5'9" to 6'11".
The data clusters between 6'5" and 6'9".
The shortest player's height (5'9") varies considerably from the mean height of 6'6".

2. **(a)** Mean number of hours:
   (0 + 0 + 1 + 1 + 1 + 1 + 2 + 2 + 2 + 2 + 2 + 3 + 3 + 3 + 4 + 4 + 4 + 5 + 5 + 6) hours ÷ 20 = 2.55 hours

   The mean number of hours worked was 2.55.

   (b)

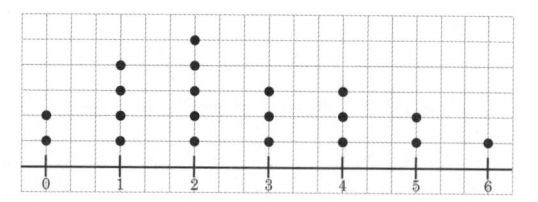

(b)

Ages (in years)	Tally	Frequency
0 – 10	𝍷𝍷𝍷𝍷𝍷𝍷 IIII	34
11 – 20	𝍷𝍷𝍷 I	16
21 – 30	𝍷𝍷 I	11
31 – 40	𝍷𝍷 I	11
41 – 50	IIII	4
51 – 60	III	3
61 – 70	𝍷	5

(c) Histogram of Age (in years) vs Frequency

Practice

3. **(a)** There are 11 students in the group.
   (b) Mean number of words:
   (10 + 10 + 12 + 13 + 13 + 13 + 15 + 16 + 29 + 36 + 40) ÷ 11
   = 207 ÷ 11 = 18.8

   The mean number is 18.8.

   (c) 10 10 12 13 13 ⓵3 15 16 29 36 40
   The median is 13.
   (d) $\frac{8}{11}$ of the students.
   (e) The outliers 29, 36, and 40 skew the mean so median is a better measure of center.

13.2B Histograms

Basics

4. **(a)** Answers will vary.
   Drawing a histogram is better than drawing a dot plot to represent this data due to the large size of this data set.

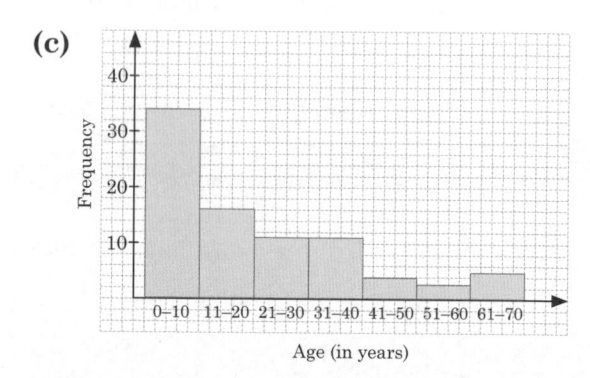

(d) $\frac{16}{84} = \frac{4}{21}$

$\frac{4}{21}$ of the people were between eleven and twenty years old.

(e) $\frac{12}{84} \times 100\% = 14.3\%$

14.3% of the people were at least forty-one years old.

Practice

5. **(a)**

Number of Steps (s)	Tally	Frequency
0 ≤ s ≤ 2,000		0
2,000 ≤ s ≤ 4,000	IIII	4
4,000 ≤ s ≤ 6,000	IIII	4
6,000 ≤ s ≤ 8,000	𝍷	5
8,000 ≤ s ≤ 10,000	II	2
10,000 ≤ s ≤ 12,000	𝍷 I	6

(b)

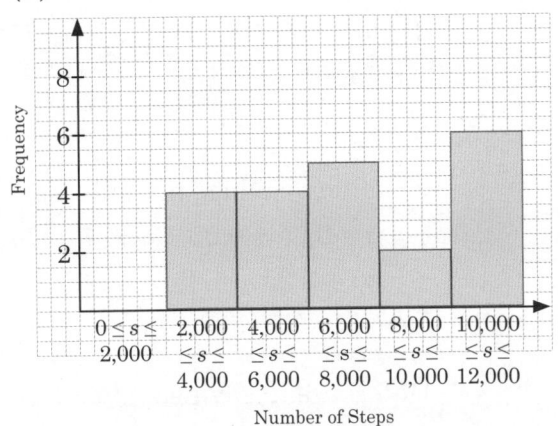

[Histogram: Frequency vs Number of Steps with intervals $0 \leq s \leq 2{,}000$, $2{,}000 \leq s \leq 4{,}000$, $4{,}000 \leq s \leq 6{,}000$, $6{,}000 \leq s \leq 8{,}000$, $8{,}000 \leq s \leq 10{,}000$, $10{,}000 \leq s \leq 12{,}000$ with frequencies approximately 0, 4, 4, 5, 2, 6]

(c) The interval between 6,000 and 8,000 steps best describes the number of steps.

6. (a) $\frac{161}{246} \times 100\% = 65.4\%$

   65.4% of the video games sold cost more than $30.

   (b) The data mounts up around the center of prices between $30 and $30.40. It is roughly symmetric and shows a bell shape.

Challenge

7. (a)

Club members	Countries visited
$\frac{1}{40} \times 120 = 3$	6
$\frac{1}{40} \times 120 = 3$	7
$\frac{1}{30} \times 120 = 4$	8
$\frac{1}{10} \times 120 = 12$	9
$\frac{7}{60} \times 120 = 14$	11
$\frac{1}{6} \times 120 = 20$	12
$\frac{7}{30} \times 120 = 28$	13
$\frac{3}{20} \times 120 = 18$	14
$\frac{1}{15} \times 120 = 8$	15
$\frac{1}{12} \times 120 = 10$	16

(b)

Number of countries	Tally	Frequency
6 – 7	|||||		6
8 – 9	||||| ||||| |||||		16
10 – 11	||||| ||||| ||||	14
12 – 13	||||| ||||| ||||| ||| ||||| ||||| ||||| |||||	48
14 – 15	||||| ||||| ||||| ||||| |||||		26
16 – 17	||||| |||||	10

(c)

[Histogram: Frequency vs Number of Countries with intervals 6–7, 8–9, 10–11, 12–13, 14–15, 16–17 with frequencies 6, 16, 14, 48, 26, 10]

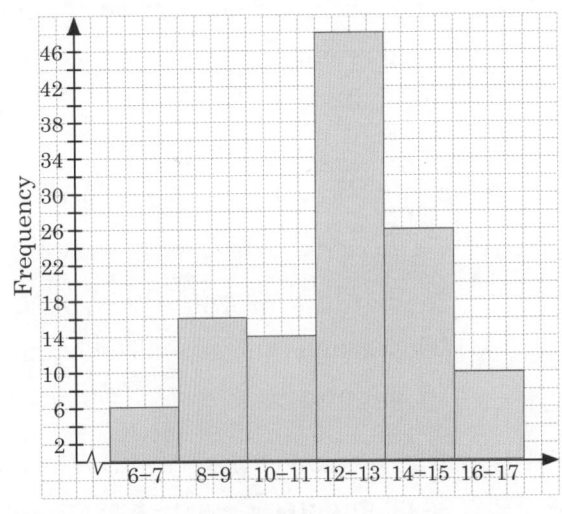

(d) Answers will vary.
The data mount up around a center of about 12-13 countries.

(e) Answers will vary.
Based on this data, the travel agent should plan to have the cruise visit at least twelve countries. 72% of the 120 club members visited at least that number of countries when on a cruise.

Chapter 13 DISPLAYING AND COMPARING DATA

Measures of Variability and Box Plots
13.3A Range

Basics

1. **(a)** Week 1
   Mean:
   $(7 + 10 + 25 + 20 + 35 + 5 + 30) \div 7$
   $= \frac{132}{7} = 18\frac{6}{7}$

   Median: 5 7 10 ⓞ 25 30 35 (20 circled)
   The mean of the first set of data is $18\frac{6}{7}$ minutes.

   The median of the first set of data is 20 minutes.

   Week 2
   Mean:
   $(15 + 18 + 20 + 15 + 25 + 26 + 28) \div 7$
   $= \frac{147}{7} = 21$

   Median: 15 15 18 ⓞ 25 26 28 (20 circled)
   The mean of the second set of data is 21 minutes.

   The median of the second set of data is 20 minutes.

   (b) Although median of both sets of data are the same, Kawai's guitar teacher would be happier to see the data for week 2. In week 1, Kawai's total practice time was 132 minutes, and practice times ranged from 5 minutes to 35 minutes. In week 2, his total practice times was 147 minutes and ranged from 15 minutes to 28 minutes.

   Range of data for week 1:
   35 min − 5 min = 30 min
   Range of data for week 2:
   28 min − 15 min = 13 min
   Kawai was more consistent in week 2.

2. **(a)** Holly's savings (in dollars):
   $(3.25 + 5.00 + 8.80 + 2.05 + 5.25 + 9.75 + 1.50 + 7.55) = 43.15$
   Holly saved $43.15.

   Jasmine's savings (in dollars):
   $(4.95 + 5.25 + 6.05 + 4.75 + 5.70 + 5.25 + 5.25 + 5.85) = 43.05$
   Jasmine saved $43.05.

   Difference in savings (in dollars):
   $43.15 − 43.05 = 0.10$
   Holly saved $0.10 more than Jasmine.

   (b) Range of Holly's savings:
   $\$9.75 − \$1.50 = \$8.25$

   (c) Range of Jasmine's savings:
   $\$6.05 − \$4.75 = \$1.30$

13.3B Mean Absolute Deviation

Basics

3. **(a)** Mean:
   $\frac{1 + 3 + 4 + 5 + 5 + 6}{6}$
   $= 4$

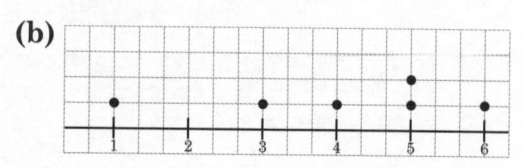

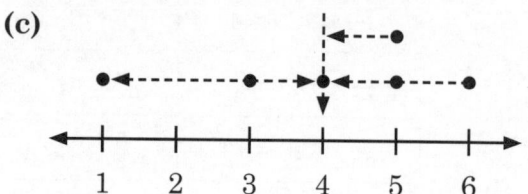

   (b)

   (c)

Chapter 13 DISPLAYING AND COMPARING DATA

(d)

Numbers in data set	Absolute Deviation (distance from the mean, 4)
1	$\|1-4\|=3$
3	$\|3-4\|=1$
4	$\|4-4\|=0$
5	$\|5-4\|=1$
5	$\|5-4\|=1$
6	$\|6-4\|=2$
	Sum = 8

(e) MAD = $\frac{8}{6}$
= 1.33

Practice

4. (a) Mean:

$(12 + 21 + 39 + 13 + 41 + 36 + 49 + 37 + 42 + 45) \div 10$

$= 335 \div 10$

$= 33.5$

The mean number of clicks is 33.5.

(b)

Number of clicks	Absolute Deviation (distance from the mean 33.5)
12	$\|12-33.5\|=21.5$
21	$\|21-33.5\|=12.5$
39	$\|39-33.5\|=5.5$
13	$\|13-33.5\|=20.5$
41	$\|41-33.5\|=7.5$
36	$\|36-33.5\|=2.5$
49	$\|49-33.5\|=15.5$
37	$\|37-33.5\|=3.5$
42	$\|42-33.5\|=8.5$
45	$\|45-33.5\|=11.5$
	Sum = 109

Mean absolute deviation of the number of clicks:

$\frac{109}{10}$ = 10.9 clicks

The number of clicks, on average, differ by 10.9 clicks from the mean of 33.5 clicks.

(c)

[dot plot from 12 to 52]

(d) The mean is not a good indicator of a typical number of clicks. There is a lot of variability in this data and the range is 37 clicks. 70% of the data points are greater than the mean.

5. (a) Imani's data

Mean number of pages:

$= (24 + 21 + 25 + 24 + 21 + 20 + 19) \div 7$

$= 154 \div 7$

$= 22$

Logan's data

Mean number of pages:

$= (9 + 10 + 18 + 16 + 20 + 40 + 41) \div 7$

$= 154 \div 7$

$= 22$

(b)

Imani's number of pages

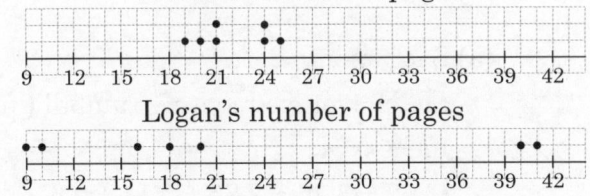

[dot plot from 9 to 42]

Logan's number of pages

[dot plot from 9 to 42]

(c) The mean of Imani's distribution would give a better indicator of a typical value. Most of the number of pages in Imani's data is closely clustered to the mean of 22. Logan's data has higher variability than Imani's.

(d) Imani's data

Absolute deviations from the mean:

$2 + 1 + 3 + 2 + 1 + 2 + 3$

$= 14$

$\text{MAD} = \frac{14}{7}$

$= 2$

Logan's data

Absolute deviations from the mean:

$13 + 12 + 4 + 6 + 2 + 18 + 19$

$= 74$

$\text{MAD} = \frac{74}{7}$

$= 10.57$

(e) Answers will vary.

The MAD values in (d) confirm the answer in (c). On average, Logan's data values are 10.57 pages away from the mean of 22, while Imani's data values are only 2 pages away from the mean.

Therefore, the mean is a better indicator of the typical number of pages read by Imani than by Logan.

13.3C Interquartile Range

Basics

6. **(a)** Sacramento data

    Mean average monthly rainfall (in inches):

    $(3.62 + 3.46 + 2.76 + 1.14 + 0.67 + 0.2 + 0.04 + 0.04 + 0.28 + 0.94 + 2.09 + 3.17) \div 12$

    $= \frac{18.41}{12} = 1.53$

    San Francisco data

    Mean average monthly rainfall (in inches):

    $(4.49 + 4.45 + 3.27 + 1.46 + 0.71 + 0.16 + 0.06 + 0.08 + 0.2 + 1.1 + 3.15 + 4.57) \div 12$

    $= \frac{23.7}{12} = 1.98$

    Bakersfield data

    Mean average monthly rainfall (in inches):

    $(1.14 + 1.22 + 1.22 + 0.51 + 0.2 + 0.08 + 0 + 0.04 + 0.08 + 0.31 + 0.63 + 1.02) \div 12$

    $= \frac{6.45}{12} = 0.54$

    (b) Sacramento data

    Median:

    0.04 0.04 0.2 0.28 0.67 (0.94 1.14)
    2.09 2.76 3.17 3.46 3.62

    $(1.14 + 0.94) \div 2 = 2.08 \div 2 = 1.04$

    San Francisco data

    Median:

    0.06 0.08 0.16 0.2 0.71 (1.1 1.46)
    3.15 3.27 4.45 4.49 4.57

    $(1.1 + 1.46) \div 2 = 2.56 \div 2 = 1.28$

    Bakersfield data

    Median:

    0 0.04 0.08 0.08 0.2 (0.31 0.51)
    0.63 1.02 1.14 1.22 1.22

    $(0.31 + 0.51) \div 2 = 0.82 \div 2 = 0.41$

(c) Sacramento data: Lower quartile and upper quartile

0.04
0.04
0.2
0.28 ← Q_1
0.67
0.94
1.14 ← Q_2
2.09
2.76
3.17 ← Q_3
3.46
3.62

Lower quartile,
$$Q_1 = \frac{(0.2 + 0.28)}{2} = 0.24$$
Upper quartile,
$$Q_3 = \frac{(2.76 + 3.17)}{2} = 2.97$$

San Francisco data: Lower quartile and upper quartile

0.06
0.08
0.16
0.2 ← Q_1
0.71
1.1
1.46 ← Q_2
3.15
3.27
4.45 ← Q_3
4.49
4.57

Lower quartile,
$$Q_1 = \frac{(0.16 + 0.2)}{2} = 0.18$$
Upper quartile,
$$Q_3 = \frac{(3.27 + 4.45)}{2} = 3.86$$

Bakersfield data
Lower quartile and upper quartile:

0
0.04
0.08
0.08 ← Q_1
0.2
0.31
0.51 ← Q_2
0.63
1.02
1.14 ← Q_3
1.22
1.22

Lower quartile, $Q_1 = 0.08$
Upper quartile,
$$Q_3 = \frac{(1.02 + 1.14)}{2} = 1.08$$

(d) Sacramento data
Interquartile range,
IQR = $Q_3 - Q_1$
    = 2.97 − 0.24
    = 2.73

San Francisco data
Interquartile range,
IQR = $Q_3 - Q_1$
    = 3.86 − 0.18
    = 3.68

Bakersfield data
Interquartile range,
IQR = $Q_3 - Q_1$
    = 1.08 − 0.08
    = 1

7. Sacramento data
   MIN = 0.04 in
   Lower quartile, Q_1 = 0.24 in
   Median, Q_2 = 1.04 in
   Upper quartile, Q_3 = 2.97 in
   MAX = 3.62 in

   San Francisco data
   MIN = 0.06 in
   Lower quartile, Q_1 = 0.18 in
   Median, Q_2 = 1.28 in
   Upper quartile, Q_3 = 3.86 in
   MAX = 4.57 in

   Bakersfield data
   MIN = 0 in
   Lower quartile, Q_1 = 0.08 in
   Median, Q_2 = 0.41 in
   Upper quartile, Q_3 = 1.08 in
   MAX = 1.22 in

13.3D Box Plot

Basics

8. (a) The data values organized from smallest to largest, in minutes:
   0
   0
   10
   10
   10 ← Q_1
   15
   20
   22
   25 ← Q_2
   27
   30
   30
   30 ← Q_3
   40
   45
   60

   MIN = 0 min
   Lower quartile, Q_1 = 10 min
   Median, $Q_2 = \dfrac{(22 + 25) \text{ min}}{2}$
   = 23.5 min
   Upper quartile, Q_3 = 30 min
   MAX = 60 min

   (b)

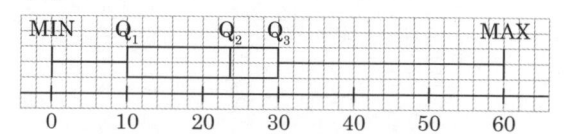

   $$\begin{aligned} \text{IQR} &= Q_3 - Q_1 \\ &= (30 - 10) \text{ min} \\ &= 20 \text{ min} \end{aligned}$$

   The number of minutes spent playing video games varies from 0 minutes to 60 minutes. The IQR is 20 minutes. The middle 50% of the number of minutes spent are between 10 minutes and 30 minutes.

Challenge

9. Answers will vary. The sum of each set of data must be 70.